The AAU REPORTS are a continuation of the Reports from the Botanical Institute, University of Aarhus. The series publishes original research undertaken by staff members and associated research workers. In addition, the series includes proceedings of symposia arranged by staff members.

The issues, continuously numbered, appear at irregular intervals and usually contain only one contribution.

The *AAU Reports* are offered in exchange to botanical libraries on request.

Editorial address and exchange request:
Department of Systematic Botany - Institute of Biological Sciences
Aarhus University
Nordlandsvej 68
DK-8240 Risskov
DENMARK
Phone (+45) 8621 0677 Fax (+45) 8621 1891

Distributed by
AARHUS UNIVERSITY PRESS
Aarhus University
DK-8000 Aarhus C, Denmark
Phone (+45) 8619 7033 Fax (+45) 8619 8433

Price 78 DKK (13 USD).
Residents of the EU should add 25% VAT.

Report editor: Henrik Balslev

Cover design: *Pireella pohlii*

Printed in Denmark by
Institute of Mathematics
University of Aarhus

ISSN 0904-6453
ISBN 87-87600-62-5

The Mosses of Amazonian Ecuador

by Steven P. Churchill

1994

AAU Reports 35
Department of Systematic Botany, University of Aarhus

Contents

Author

Steven P. Churchill. *Born 1948. B.S. from the University of Nebraska; M.A. from the University of Kansas; Ph.D. from The New York Botanical Garden and City University of New York in 1988. Visiting Scientist, Botanical Institute, University of Aarhus, 1988–1989. Research Associate (1989–1993), Assistant Curator (1993–present) at The New York Botanical Garden, Bronx, NY 10458-5126, U.S.A. Tel (+1) 718-817-8641, fax (+1) 718-562-6780, email: schurchill@nybg.org.*

Acknowledgments

I thank the Danish Research Academy for supporting the author's stay at the University of Aarhus which made this present effort possible, and the Danish Natural Science Research Council which made possible field work in Amazonas of Ecuador through a grant to H. Balslev. It has been an enjoyable and a stimulating experience to work with the staff, technical assistants, and students of the Botanical Institute and herbarium while at the University of Aarhus. I thank the curators of AAU, BM, GB, NY, and S for use or loan of specimens during the course of this study. Dr. Gert Morgensen assisted with obtaining various literature during the course of this study. Sven Fransén kindly contributed data concerning his and other staff collections from Amazonas contained at GB, and also provided information on the genus *Bartramia*; likewise Dr. Inés Sastre-De Jesús provided information on the Neckeraceae and Thamnobryaceae. I thank both Drs. William Buck and Carmen Ulloa U., for suffering at my request to review the manuscript, both provided many useful comments and suggestions for which I am most grateful. Ms. Alba Luz Arbeláez provided not only the Spanish translation but assisted with various aspects in the preparation of the manuscript. Finally, my thanks to Dr. Henrik Balslev, for his friendship, encouragement, and assistance during the course of this work.

Lastly, I wish to express my deep appreciation to the late Dr. William C. Steere. My early development and interest in Ecuador and the Neotropics owes much to his kindness and willingness to assist not only with my initial collections, but also freely provided comments and opinions on various aspects of neotropical bryology, including personal tales of adventure and misadventure in the Andes; this work is merely an extension of his inspiration.

Acknowledgments

I thank the Danish Research Academy for supporting the author's stay at the University of Aarhus which made this present effort possible, and the Danish Natural Science Research Council which made possible field work in Amazonas of Ecuador through a grant to H. Balslev. It has been an enjoyable and a stimulating experience to work with the staff, technical assistants, and students of the Botanical Institute and herbarium while at the University of Aarhus. I thank the curators of AAU, BM, GB, NY, and S for use or loan of specimens during the course of this study. Dr. Gert Morgensen assisted with obtaining various literature during the course of this study. Sven Fransén kindly contributed data concerning his and other staff collections from Amazonas contained at GB, and also provided information on the genus *Bartramia*; likewise Dr. Inés Sastre-De Jesús provided information on the Neckeraceae and Thamnobryaceae. I thank both Drs. William Buck and Carmen Ulloa U., for suffering at my request to review the manuscript, both provided many useful comments and suggestions for which I am most grateful. Ms. Alba Luz Arbeláez provided not only the Spanish translation but assisted with various aspects in the preparation of the manuscript. Finally, my thanks to Dr. Henrik Balslev, for his friendship, encouragement, and assistance during the course of this work.

Lastly, I wish to express my deep appreciation to the late Dr. William C. Steere. My early development and interest in Ecuador and the Neotropics owes much to his kindness and willingness to assist not only with my initial collections, but also freely provided comments and opinions on various aspects of neotropical bryology, including personal tales of adventure and misadventure in the Andes; this work is merely an extension of his inspiration.

1. INTRODUCTION

This present treatment is a guide to the mosses known from and some likely to be found in Ecuadorean Amazonas. The purpose of this florula is to stimulate further field work by those residing or working in Ecuador by providing a means to identify collections and thus add to our understanding of the diversity and distribution of the mosses of this vast and interesting region of Ecuador. Future work should contribute not only better distribution data, but also add considerable new additions to the mosses of the eastern lowlands and new records for Ecuador. Furthermore, it is hoped that such a guide will promote the use of mosses in other ways, such as for inventory and ecological studies. Although the Amazonas mosses of Ecuador are still undercollected, our knowledge is far better than exists for comparable areas in Bolivia, Colombia and Peru which have received little attention in terms of field collections. This is largely due to the excellent cooperative efforts between the P. Universidad Católica in Quito and the participating Nordic countries of Denmark and Sweden to study Ecuador's flora. This guide should also be somewhat useful in the Pacific lowlands of western Ecuador, and also in the Amazonas of Colombia and Peru.

Bryophytes are a common vegetational element in the Neotropics. However, in general their taxonomy and ecology have received little attention and remain poorly known. Some bryologists have suggested, with some justification, that bryophyte floras for tropical regions are premature, introducing errors such as the mis-use of names or describing new superfluous taxa. The proposed alternative is to await completion of regional or worldwide monographic studies. The serious flaw in this reasoning is that when such revisions are completed, possibly in the early part of the next century, a large part of the existing vegetation will have disappeared. Nor does such reasoning provide those residing in tropical countries (and we are speaking of several generations) access to such information, not only in terms of taxonomic studies, but no opportunity to conduct research in other areas of comparative biology. Floras are necessary, perfect or not. It has been suggested that florulas of local or regional areas can possibly help to rapidly promote an understanding of the flora as a first step (Gentry, 1978; Prance, 1984), and at the same time make available information that can be used by others in comparative biology. Few such examples exist for the bryophytes of tropical South America, however an excellent example is that by Griffin (1979) for the Manaus region of Brazil. This present treatment is modeled, in part, after the florula by Griffin. Here, I have tried to provide in a synoptical way, general information, keys and glossary in Spanish. Latin America is in critical need of such treatments. The present status and needs in bryology for Latin America have recently been discussed and reviewed by Delgadillo M. (1982), Griffin and Gradstein (1982), Matteri (1985), and Buck and Thiers (1989). It is hoped that in the future greater interest will be generated in an effort to study mosses, particularly by those residing in Ecuador.

Finally, I must point out that this project initially began at the Botanical Institute with the idea of producing keys only to the mosses of the Añangu area in Napo Province. From that grew the present treatment. Portions of this work reflect my own interest or knowledge of various groups as well as my ignorance of others. As such, it is far from perfect but I hope that it contributes some knowledge about the mosses of Ecuadorean Amazonas.

Geographical Location and Vegetation

The eastern tropical lowlands (Fig. 1), that form a part of the upper or western Amazon basin, is the largest vegetation zone in Ecuador with an estimated area of 71,000 square km, nearly one-third of Ecuador, defined by the elevational boundary of 600 m and below following Balslev and Renner (1989), and Renner et al. (1990). This region of Ecuador has in past times been referred to as the Oriente and more recently as the Amazonas or the eastern lowlands. This area contains the larger part of the provinces of Morona-Santiago, Pastaza, and Napo, the latter now divided with Sucumbíos Province laying approximately north of the Río Napo and Napo proper laying to the south. The physiognomy of this region is mostly flat plains with numerous large and small tributaries that drain from the eastern Andean slopes to the Amazon River. Some of the major river systems include the Aguarico, Bobonaza, Curaray, Napo, and Pastaza. The aseasonal climate of this region is characterized by high precipition, 2000–6000 mm per year, and constant temperates above 24°C throughout the year (Balslev, 1988).

The forests of the eastern lowlands are characterized by a dense canopy, with average heights of about 30 m above ground level. Much of the vegetation is divided between forested non-flooded areas, terra firme (Fig. 1), and forested floodplain areas that are

Figure 1. Ecuadorean Amazonas forest. (Photo: H. Balslev)

often inundated (Fig. 2). A rich diversity is found among trees, epiphytes and lianas. For the vascular plants, including ferns, it has been estimated that the Amazonas of Ecuador contains some 3300 species of which about 40% are trees (Balslev and Renner, 1989). The 13 most important families of vascular plants containing approximately 46% of the species are: Orchidaceae, Rubiaceae, Melastomataceae, Leguminosae, Moraceae, Arecaceae, Annonaceae, Solanaceae, Lauraceae, Myrtaceae, Piperaceae, Euphorbiaceae, and Araceae (Balslev and Renner, 1989).

Some estimate of species richness among trees (≤ 10 cm dbh) is provided in a series of studies at Añangu located along the Río Napo in Napo Province. As summarized by Balslev and Renner (1989), two transects, one each from an inundated and *terra firme* forest, and one plot study from a *terra firme* forest, were confined to an area about 3.5 × 0.5 kilometers. These three sample sites at Añangu yielded 394 species distributed among 1958 trees. Furthermore, the percent of shared tree species between the three sample sites were equal to, or less than 30%, suggesting that the Añangu area is probably far more species rich than presently indicated.

A summary of the flowering plants of the Ecuadorean lowlands has recently been provided by Renner *et al.* (1990) that gives more detailed information with regards to the physical aspects and vegetation of this region.

Amazonas Mosses

This present study treats 119 species of mosses distributed among 62 genera and 27 families. The five largest Ecuadorean Amazonas families are: Callicostaceae (25 spp.), Fissidentaceae (11 spp.), Calymperaceae (10 spp.), Hypnaceae (9 spp.), and Sematophyllaceae (8 spp.). The number of mosses for any restricted site of a few hectars in Amazonas apparently does not reach more than about 50 species and often is much lower than that number (Churchill *et al.*, 1992).

The number of moss species expected for Ecuador is presently thought to be between 800–900 (874 species presently known, see Appendix 1). Thus, the Amazonas mosses represent only about 12 to 14% of the total Ecuadorean moss flora. A study of the elevational distribution in Colombia (Churchill, 1991b) demonstrated that approximately 10% of the total known moss flora occured in the lowlands (600 m or below), and that the greatest taxonomic

Figure 2. Inundated floodplain forest in Amazonas. (Photo: H. Balslev)

diversity, about 50%, was found in the high montane and transitional montane-subpáramo zones (2600–3300 m). A similar trend for the mosses is predictable for Ecuador.

Factors influencing the low species richness and biomass of mosses in the wet lowland tropical forests, as compared to the signficant number of species and greater increase of biomass in the montane forests have been discussed by various authors in recent times (*e.g.*, Pócs, 1982; Richards, 1984). These authors have suggested a combination of factors related to increase of species richness and biomass: decreasing temperatures, increasing humidity, and possibly more exposure. Recently it has been demonstrated that net photosynthesis in tropical lowland mosses is insufficent due to high temperatures and low light intensities, thus possibly accounting for the relatively low biomass and low number of species (Frahm, 1987b).

Habitats and Growth Forms

The majority of Amazonas mosses are found as epiphytes, occurring on bark (corticolous or ramicolous) or leaves (epiphyllous) of trees, treelets, shrubs and lianas. Less frequently mosses are found on logs (lignicolous). Fewer still are found on rocks (saxicolous) or soil (terricolous) and then either associated with exposed roots from fallen trees, such as the genus *Fissidens*, or on exposed stream or river banks, or cut banks along roads or trails. A few are also semi-aquatics or aquatics found on rocks in streams.

Mosses exhibit a wide range of growth forms. The pleurocarpous mosses, those that produce the sporophyte laterally on stems and generally exhibit secondary stems that are procumbent, pendent, or erect and frondose or dendroid, account for ca. 65% of the species in Amazonas. Most of these occur as epiphytes. Such pendent or pendulous forms are found in the Meteoriaceae, while others form spreading or ascending mats such as in the Hypnaceae or Sematophyllaceae, which are also found on logs. Still others exhibit dendroid or frondose forms, such as in some members of the Callicostaceae, Leptodontaceae and Pterobryaceae.

The acrocarpous mosses, those that produce the sporophyte terminally on erect stems, form loose or dense tufts or cushions, rarely occurring gregariously, account for ca. 45% of the Amazonas species. Most acrocarpous mosses are commonly found as epiphytes, less often on logs, such as in the Calymperaceae and Leucobryaceae. Some members of the Bryaceae, Dicranaceae and Pottiaceae are more frequent on exposed clay banks along waterways or cutbanks along trails or roads.

Growth forms also contribute to our understanding of diversity. This subject has been reviewed and discussed recently by Mägdefrau (1982) and Richards (1984) for bryophytes. The Ecuadorean Amazonas mosses were grouped under five major growth forms modified from that given in Table 1; definitions of these various growth forms are given in the table. This present effort was simply to establish basic patterns of growth forms. Turf and/or cushion forming mosses (Table 1: A1 & A2) account for 39% of the taxa, while mat forming mosses (Table 1: A3) were second with 34%; combined these two contributed 73% of the growth forms found in Amazonas. A smaller percent of the growth forms include feather (11%; Table 1: B3), pendulous (10%; Table 1: B4), and dendroid (6%; Table 1: B2); combined these three contributing only 27% of the growth forms in Amazonas mosses. This illustrates the rather wide range of growth forms to be found in Amazonas. With

---

Table 1. Growth forms of tropical mosses adapted and modified from Richards (1984), with examples of Amazonas taxa.

A. Social forms. Leafy shoots aggregated.
    1. Cushions (shoots mainly erect and radiating to form dome-shaped masses; large cushions = > 5 cm diameter, small cushions = < 5 cm diameter); *e.g. Leucobryum.*
    2. Turfs (shoots upright, ±parallel).
        a. Tall turfs (> 2 cm high); *e.g. Holomitrium, Leucoloma,* some *Octoblepharum* spp.
        b. Tall turfs with divergent or creeping branches; *e.g. Acroporium,* Orthotrichaceae.
        c. Short turfs (< 2 cm high); *e.g. Barbula, Bryum,* some *Octoblepharum.*
        d. Open turfs (shoots ±separated) *e. g. Fissidens.*
    3. Mats (primary stems creeping horizontally over substratum, lateral branches erect or parallel, usually forming closely interwoven mats).
        a. Rough mats (main shoots adhering to substratum, with abundant short erect laterals); *e.g., Leucomium, Sematophyllum.*
        b. Smooth mats (branches in same plane as main shoot); *e.g.,* many Callicostaceae (including some *Lepidopilum* spp.), Hypnaceae, *Taxithelium.*
        c. Wefts; *e.g., Thuidium.*
B. Solitary forms. Leafy shoots not closely aggregated.
    1. Protonemal mosses (protonema persistent, leafly shoots minute, scattered); not represented in Amazonas.
    2. Dendroid, not dorsiventral; *e.g. Pireella, Porotrichum, Pseudocryphaea.*
    3. Feather forms (secondary stems ±pinnately branched, dorsiventral); *e.g., Callicosta,* some *Lepidopilum* ssp., *Neckeropsis.*
    4. Pendent forms (secondary stems and branches usually long and pendulous); *e.g.,* Meteoriaceae, *Hildebrandtiella* and *Orthostichopsis.*

greater study, further subdivision could be made than the five groups recognized above but more detailed field studies are needed. A number of species could be grouped under one of several catagories recongized in Table 1. For example, *Hydropogon* can form mats but more often is pendent, and this is also true of several genera, such as in the Meteoriaceae.

Floristic Composition and Comparison

A comparison of neotropical wet lowland forest mosses from Ecuadorean Amazonas and three previous studies, Guyana (Richards, 1934), Manaus (Griffin, 1979), and West Suriname (Florschütz-de Waard and Bekker, 1987), is presented in Table 2. It can be noted that a number of species are shared between these three areas and the Amazonas of Ecuador. These species, common to the four compared areas, are generally widespread in the neotropical wet lowlands, examples included: *Callicosta bipinnata, Callicostella pallida, Crossomitrium patrisiae, Leucobryum martianum, Leucomium strumosum, Neckeropsis disticha, N. undulata, Octoblepharum albidum, Pilosium chlorophyllum, Sematophyllum subpinnatum, S. subsimplex, Taxithelium planum, Thuidium involvens, Vesicularia vesicularis,* and *Zelometeorium patulum.* The percent of shared species between Ecuadorean Amazonas and the three areas are: Guyana (28%), Manaus (35%), and West Suriname (35%). However, there are some conspicuous differences between Ecuadorean Amazonas and the other areas. Considerably more species of the genera *Fissidens* (11) and *Lepidopilum* (10) are found in Amazonas, whereas five, four, and six species of *Fissidens* are recorded from Guyana, Manaus, and West Suriname respectively, and three *Lepidopilum* species each from West Suriname and Guyana, and none from Manaus. Furthermore, *Cyclodictyon* and *Porotrichum* were not recorded in the other three areas. In contrast, *Calymperes* in Amazonas is represented by only three species, whereas in West Suriname and Guyana there are five and three species respectively, and in Manaus nine species. Even more conspicuous are the greater number of species in the genus *Syrrhopodon* from Manaus with 14 species in contrast to Amazonas with seven, and West Suriname and Guyana with five each, the latter two areas share all five species with Amazonas. Similarly, *Octoblepharum* is better represented in Manaus (7 spp.) than the other three areas which record three or less species. Less contrasting differences exist between each of the four areas but generally involve one or two species. However there is enough dissimilarity to suggest that the tropical wet lowlands are not merely composed of widespread species. Some of the differences may reflect an artifact of collecting, however all four areas have, at least in part, been studied by bryologists. At least one factor contributing to the distinct composition of taxa concerns a historical component. The refugium theory, which has endeavored to explain species diversity and distribution, has played a significant part in the interpretation of Amazonian vegetational history in recent years (cf., Whitmore and Prance, 1987). This theory has more recently been criticized from various fields of investigation (for an overview see Gentry, 1989). Recent studies by Colinvaux and associates (see Colinvaux, 1989, and Bush et al. 1990) have suggested that the lowlands, at least in the upper Amazon basin, were not as arid as predicted earlier for the Pleistocene period and that the elevational lowering during this period may have allowed for a mixing of upland and lowland taxa in areas adjacent to the Andes. Furthermore, Salo and Räsänen (1989) have shown that the western Amazon region is a highly dynamic system of changing landscapes related to short or long term geomorphic processes. Probably many of these factors have influenced the patterns observed, and may account for some of the differences suggested in the distribution of lowland mosses in Amazonian Ecuador, at least at the species and possibly at the generic level.

History of Collecting and Publications

Few naturalists and even fewer bryologists have collected in Ecuadorean Amazonas and thus few publications deal with the mosses of this region. Two publications (De Notaris, 1859; Bartram, 1934) that refer to the Río Napo in their titles actually concern mostly montane and páramo mosses (see Steere, 1988 and Churchill et al., 1992). Bartram's (1934) title, "Mosses of the River Napo, Ecuador" contains an account of Manuel Villavicencio's collections from Ecuador made "along the banks of the River Napo" in 1869. Of the 83 species enumerated by Bartram, 10 could have actually been made in Amazonas, but all 10 species are also known either from the low montane or premontane forests. However, Villavicencio apparently did make at least two collections (*Isopterygium tenerum* and *Octoblepharum albidum*) from Amazonas, at Puerto del Napo during May, 1847 (see Steere, 1948).

Table 2. Comparison of the number of species unique and shared among families and genera from tropical wet lowland forests of Ecuadorean Amazonas (presented here), with West Suriname (Florschütz-de Waard & Bekker, 1987), Guyana, Moraballi Creek (Richards, 1934), and Manaus region (Griffin, 1979). Only species of the wet or moist forests calculated for the Guyana and West Suriname areas. Numbers in parentheses are the number of species shared with Ecuadorean Amazonas.

	Ecuadorean Amazonas	West Suriname	Guyana	Manaus
Bartramiaceae				
Philonotis	3	0	0	1 (1)
Brachytheciaceae				
Oxyrrhynchium	1	0	0	0
Bryaceae				
Bryum	3	0	0	2 (2)
Callicostaceae				
Brymela	1	0	1	1
Callicosta	3	2 (1)	2 (1)	2 (1)
Callicostella	3	4 (2)	2 (1)	1 (1)
Crossomitrium	2	1 (1)	1 (1)	1 (1)
Cyclodictyon	4	0	0	0
Hypnella	1	0	1 (1)	0
Lepidopilidium	0	0	1	0
Lepidopilum	10	3 (3)	3 (3)	0
Pilotrichidium	1	0	0	0
Thamniopsis	0	0	2	0
Calymperaceae				
Calymperes	3	5 (3)	3 (2)	9 (3)
Syrrhopodon	7	5 (5)	5 (5)	14 (7)
Daltoniaceae				
Leskeodon	1	0	0	0
Dicranaceae				
Campylopus	0	0	0	3
Dicranella	1	1 (1)	0	1 (1)
Holomitrium	1	1 (1)	1 (1)	0
Leucoloma	1	0	0	0
Microdus	1	0	0	0
Trematodon	1	0	0	0
Fissidentaceae				
Fissidens	11	6 (3)	5 (2)	4 (3)
Hydropogonaceae				
Hydropogon	1	0	0	0
Hypnaceae				
Chryso-hypnum	1	0	0	0
Ectropothecium	2	1 (1)	1 (1)	0
Isopterygium	1	1 (1)	0	2 (1)

Mittenothamnium	1	0	0	0
Phyllodon	1	0	0	0
Rhacopilopsis	1	0	1 (1)	1 (1)
Taxiphyllum	1	0	0	0
Vesicularia	1	1 (1)	1 (1)	1 (1)
Leptodontaceae				
Pseudocryphaea	1	0	0	0
Leucobryaceae				
Leucobryum	1	1 (1)	2 (1)	2 (1)
Leucophanes	1	1 (1)	0	0
Octoblepharum	3	2 (1)	1 (1)	7 (3)
Leucomiaceae				
Leucomium	1	1 (1)	1 (1)	1 (1)
Meteoriaceae				
Meteoridium	1	0	0	0
Papillaria	1	0	0	0
Pilotrichella	1	0	0	0
Squamidium	2	0	0	0
Zelometeorium	2	1 (1)	1 (1)	1 (1)
Neckeraceae				
Neckeropsis	2	2 (2)	1 (1)	2 (2)
Orthotrichaceae				
Groutiella	1	2 (1)	1	1
Macromitrium	2	1	2	2
Schlotheimia	1	2 (1)	1 (1)	0
Phyllodrepaniaceae				
Mniomalia	1	1 (1)	0	1 (1)
Phyllodrepanium	0	0	1	1
Phyllogoniaceae				
Phyllogonium	1	0	0	0
Pottiaceae				
Barbula	2	0	0	1
Dolotortula	1	0	0	0
Hyophila	1	1 (1)	0	1 (1)
Pterobryaceae				
Henicodium	1	1 (1)	0	1 (1)
Hildebrandtiella	1	0	0	0
Jaegerina	0	1	0	0
Orthostichopsis	1	1	0	0
Pireella	1	0	0	1 (1)
Racopilaceae				
Racopilium	1	0	0	0
Rhizogoniaceae				
Pyrrhobryum	1	0	0	0
Sematophyllaceae				
Acroporium	2	1 (1)	2 (2)	1 (1)
Meiothecium	0	0	0	1
Potamnium	0	1	0	0

Pterogonidium	0	0	0	1
Sematophyllum	2	2 (2)	2 (2)	2 (2)
Taxithelium	1	1 (1)	1 (1)	1 (1)
Trichosteleum	3	3 (1)	1 (1)	2 (1)
Sphagnaceae				
Sphagnum	0	0	0	1
Splachnobryaceae				
Splachnobryum	1	0	0	0
Stereophyllaceae				
Entodontopsis	0	0	0	1
Pilosium	1	1 (1)	1 (1)	1 (1)
Thamnobryaceae				
Pinnatella	1	0	0	0
Porotrichum	4	0	0	0
Thuidiaceae				
Thuidium sensu lato	3	1 (1)	0	2 (1)
Total	119	59 (42)	48 (33)	65 (42)
Total percent of shared species				
between Ecuadorean Amazonas	35%	28%	35%	

The second and most significant collections from Amazonas Ecuador were made by Richard Spruce (1817–1893) in 1857. The majority of Spruce's collections were made along Río Bobonaza, a tributary to Río Pastaza, Pastaza Province. Spruce first entered the mouth of Río Bobonaza on 6 May on his way to Canelos which he finally reached on 12 June. Mitten (1869) enumerated Spruce's moss collections, listing approximately 40 species from Ecuadorean Amazonas, most from Río Bobonaza, of which about 14 species were described as new to science. The major sets of Spruce's collections can be found at BM, E, and NY. One final comment concerning Spruce involves the actual name of the locality that these collections were gathered. The name Río Bobonaza appears as "Bombonasa" on Spruce's labeled moss collections (also *cf.* Mitten, 1869, and many revisionary studies since), and this spelling is also contained in Wallace's edited version of Spruce's (1908) travels in South America.

Nearly 70 years passed before naturalists and plant collectors made any significant collections of Amazonas mosses. Raymond Benoist (1881–1970) collected eight mosses along the Río Napo at an elevation of 500 meters in the early 1930's and these were reported on by Thériot in 1936. The Benoist collections should be found at PC. Ynés Mexía (1870–1938) made at least some collections around Gavilanes and Las Palmeras, in the vicinity of Tena in 1935. The Mexía collections can be found in several bryological herbaria such as FH or NY. The first and only catalogue prepared for the Ecuadorean mosses was by Steere (1948). Although this catalogue did not contain Steere's own collections made in the 1940's from Ecuador which were not processed at that time, the list contains an extensive account of the literature up to that time. Also a brief overview of the phytogeography as well as an account of the collectors and history regarding Ecuadorean mosses was given which is still a useful introduction. Describing Amazonas as one of the least understood regions in Ecuador, Steere's catalogue listed approximately 50 species from the region. In 1947 Gunnar Harling collected a number of mosses in southern Amazonas, around Patuca and Yurupaza in the present province of Morona-Santiago. These collections, in addition to other Ecuadorean collections made by Harling, were reported on by Crum in 1957 and included 13 species from the eastern lowlands. The Harling collections can be found at S and dulpicates at CANM. The Oxford University Expedition in 1960 conducted a comparative vegetational analysis of montane and lowland forests in eastern Ecuador which also mentioned bryophytes (Grubb et al., 1963; Grubb and Whitmore, 1966, 1967). The mosses collected during 1960 by Grubb and his collaborators were reported on by Bartram in 1964. Once again, another of Bartram's publication titles on Ecuadorean mosses was rather misleading. "Mosses of Cerro Antisana" suggests high elevational species from montane and páramo vegetational zones. Collections were, in fact, made to over 4000 m, but also down to Amazonian lowlands at ca. 400 m. Contained in Bartram's publication are approximately 24 species from the wet lowland forests near Shinguipino in Napo Province. Three new species from Amazonas of the genus *Lepidopilum* were described, all of which are now considered synonyms. The Grubb collections can be found at the BM and FH. In 1968 Lauritz Holm-Nielsen and Stig Jeppesen collected 19 species of mosses from Amazonas which were included in the report by Robinson et al. in 1971. The Holm-Nielsen and Jeppesen collections can be found at AAU and US. In 1985 Steven Churchill and Inés Sastre-De Jesús inventoried the bryophytes along Río Napo at Añangu in Napo Province. A total of 50 species of mosses were recorded (Churchill et al., 1992). The Churchill and Sastre-De Jesús collections can be found at NY, AAU and QCA. At this time, based on the above literature cited, approximately 100 species were recorded from the Amazonas of Ecuador, several which are now considered synonyms.

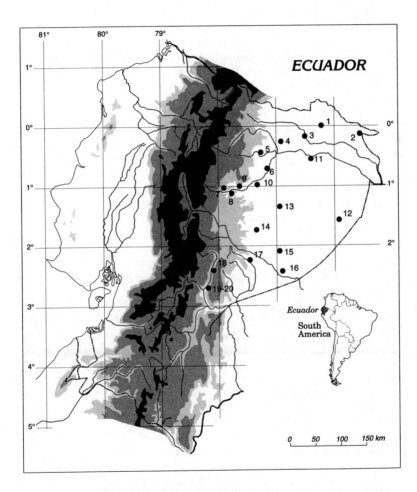

Figure 3. Some major collecting localities in Amazonas. 1. Cuyabeno, 2. Rio Güepi, 3. San Pablo de Los Secoyas, 4. Las Sachas, km 40, 5. Río Payamino, 6. Río Suno, 7. Río Pano, near Tena, 8. Shinguiipino, 9. Mishualli, 10. Yuraipa, 11. Añangu, 12. Lorocachi, 13. Curaray, 14. Río Bobonasa [Río Bombanaza], 15. Montalvo, 16. Destacamento Chiriboga & Apachi Entza, 17. Taisha, 18. Médez-Sucúa road, 19 & 20. Yurupaza & Patuca.

With the initiation of the Flora of Ecuador project in the 1960's, which has continued to the present time, numerous bryophytes have been collected by staff and students at the University of Aarhus (AAU) and the University of Göteborg (GB), as well as the P. Universidad Católica in Quito (QCA). Among these collectors for Amazonas, in addition to some of those listed above, particularly Holm-Nielsen, include: Lennart Andersson, Eduardo Azanza C., Henrik Balslev, John Brandbyge, Flavio Coello H., Sven Fransén, Jaime Jaramillo A., Bernt Løjtnant, Ulf Molau, and Benjamin Øllgaard. Many of these collections are cited in the taxonomic treatment. Many of the AAU and GB collections have been determined by William C. Steere (1908–1989).

More recently an inventory study located at Reserva Faunística Cuyabeno in the province of Sucumbíos has been initiated by the University of Aarhus. Initial bryophyte collections were made by Henrik Balslev, and in early 1990 Risto Heikkinen conducted an ecological and general survey of the bryophytes at Cuyabeno; many of these collections are included in this study.

Finally, a detailed catalogue of all plant collectors, including bryologists, has recently been compiled by Susanne Renner (1993). This provides very useful information on collectors, where collections are deposited, and publications related to collections made in Amazonas.

The present status of previous collecting in Amazonas is given in Figure 3. Napo Province has twice as many localities where mosses have been collected as compared with the provinces of Morona-Santiago and Pastaza. Sucumbíos has only two such sites. Clearly more collecting should yield further additional records for Amazonas, particularly in the southern provinces. However, much has been accomplished in recent years, primarily by vascular plant collectors who have made a major contribution to our understanding of the mosses of this region.

Collecting and Identification of Mosses

Bryophytes are possibly among the easiest plants to collect. They can be gathered and placed in paper bags or folded newspaper. They should be allowed to dry before placing in newspaper or bags if possible, or simply to leave the folded newspaper or paper bags slightly open for the plants to dry. When collecting, it is best not to mix collections though this is not always possible to avoid. As noted previously, mosses grow in a variety of habitats and on various substrates. While it is most convenient to collect material that form large tufts or mats, it should be remembered that a number of species occur as only one or a few individuals, and these may prove to be the most interesting, not the most abundant and conspicous mosses. It is NOT necessary to collect plants with sporophytes! This is particularly true in Amazonas with a limited number of species, however it is also true of 80–90% of the mosses of Ecuador which can be identified simply with the characters exhibited on the gametophytic plant.

Basic information to record is similar to that for vascular plants. The minimum information to record and to use for labels is given with the following example.

Flora of Ecuador

Callicostaceae
Lepidopilum affine C. Müller

Prov. Napo. Añangu, south shore of Río Napo, ca. 75 km east of Puerto Francisco de Orellana -"Coca" (ca. 00°32′S, 76°23′W). Tropical rain forest. 245-325 m. Moss on tree trunk ca. 1 m above ground. 29.vii.-4.viii.1985 *S. P. Churchill & I. Sastre-De Jesús No. 13798.*

If convenient, as it often is, duplicates can easily be made. It is useful to make duplicate collections for either exchange or to obtain determinations by those working on floristic or revisionary studies. The first set of collections must be deposited in one of several major herbaria in Ecuador, such as QCA. Other herbaria specializing in neotropical mosses that would be very interested in receiving bryophyte collections as a gift or for exchange include some of the following herbaria: COL, MEXU, MO, NY, PMA, S, and US.

Basic equipment necessary to identify mosses include both a stereo and a light microscope. Dissecting needles are useful, especially in which one has the end point flattened and sharp, like a spear point. To examine a collection, usually one or two individual plants are sufficent with most acrocarpous mosses or a portion of stem and branch of a pleurocarpous moss; these can be placed in water, best at boiling point and then placed on a slide. With two dissecting needles and/or razor blade, one can hold down the distal stem with one needle and with the other flattened needle point or a razor blade slightly angled, strip off the leaves. A cover glass can be placed over the leaves, adding enough water to fill in the area between the slide and cover glass. For the most part, the mosses described here can be determined on the shape and structure of the leaves. While a good stereo scope can be used to observe the general features of the gametophyte, particuarly the growth form, and sporophyte, a light microscope will provide the best results for detailed characters, particuarly of the patterns and ornamentation of leaf cells and peristome.

It is important to differentiate between mosses and hepatics, which sometimes is not easy. I usually can tell simply by the general appearence, if they are attractive, that is aesthetically pleasant to look at, they are probably mosses, however that is not a convenient guide and few hepaticologists would agree to such criteria (in fact hepatics are rather interesting and even attractive, so much so that on his return to England Richard Spruce devoted the remaining years to study and discribe his collections from the Amazon and Andes; see Thiers, 1984). For those wishing to know more about hepatics, the keys provided by Cole (1983, 1984) and Griffin (1979) are most useful. The following general characteristics may be of better help to distinguish mosses from hepatics in the field (with the aid of a handlens) and in the laborartory with a stereo scope or microscope.

Mosses	Hepatics
Gametophyte:	
Leafy, leaves mostly spirally arranged.	Leafy & thalloid, leaves 2-ranks with a small ventral leaf.
Leaves not lobed or segmented.	Leaves lobed or segmented.
Costa present (usually).	Costa absent.
Lamina cells elongate or isodiametric.	Lamina cells isodiametric.
Sporophyte:	
Seta mostly elongate, wiry & persisting.	Seta often short, if elongate, then weak & ephemeral.
Opercula & peristome present on capsules.	Opercula & peristome absent.
Elaters absent	Elaters present

2. RESUMEN

Introducción

Este trabajo que presenta una guía para el conocimiento de los musgos de la amazonía Ecuatoriana, pretende estimular el trabajo de campo y la investigación en Ecuador, al proveer los medios necesarios para el conocimiento de su diversidad, en esta vasta e interesante región del Ecuador. A pesar de que los briófitos son un elemento común de los Neotrópicos, su taxonomía y ecología son muy poco conocidas.

Las tierras bajas al este del Ecuador están definidas por el límite elevacional de 600 m ó menos, de acuerdo a Balslev (1988) y Balslev y Renner (1989). Esta región ha sido considerada como el Oriente o amazonas y comprende las provincias de Morona-Santiago, Pastaza, Napo y Sucumbíos. La vegetación de esta región es de bosque lluvioso húmedo tropical le forma parte del occidente de la Cuenca Amazónica. Se reconocen dos tipos de bosques: i) tierra firme o bosques no inundados, ii) Bosques planos a menudo inundados.

Musgos del Amazonas

Este trabajo incluye 119 especies distribuidas en 63 géneros y 27 familias. Las cinco familias más grandes en el amazonas son: Callicostaceae, Fissidentaceae, Hypnaceae, Sematophyllaceae y Meteoriaceae. Un total de cerca de 874 especies distribuidas en 233 géneros y 62 familias se encuentran en Ecuador (véase Apéndice 1), así los musgos del amazonas representan solo el 12–14 % del total de los musgos en Ecuador. El número de musgos incrementa con la elevación, alcanzando su máximo en número y abundancia, en bosque montano alto. La mayoría de los musgos del amazonas ocurren como epífitos sobre corteza u hojas, y son menos comunes sobre troncos en descomposición, suelo y rocas. Las diferentes formas de crecimiento en musgos estan presentadas en la tabla 1A. Las formas de crecimiento, aparecen simplificadas en cinco grupos, mostrando el respectivo porcentaje de cada una en el amazonas: i) Forma de racimos y/o cojines - 39%; ii) Forma de esteras - 34%; iii) Forma de pluma - 11%; iv) Formas péndulas - 10%; v) Formas dendroides - 6%.

Las colecciones más importantes del siglo 19 fueron realizadas por Richard Spruce (1818–1893) en 1857, principalmente a lo largo del Río Bobonaza. Spruce colectó cerca de 40 especies en el amazonas, de las cuales 14 fueron nuevas (Mitten, 1869). En 1948 William Spruce publicó el primer catálogo de los musgos del Ecuador basado en reportes de literatura donde, aproximadamente 50 especies fueron reportadas para el amazonas. Colecciones más recientes incluyen aquellas realizadas por Harling (Crum, 1957), Grubb (Bartram, 1964), Holm-Nielsen & Jeppesen (Robinson *et al.*, 1971) y Churchill & Sastre de Jesús (Churchill *et al.*, 1992). A partir del inicio del proyecto Flora de Ecuador, muchas colecciones han sido realizadas por miembros y estudiantes de la Universidad Católica de Quito (QCA), la Universidad de Århus (AAU) y la Universidad de Göteborg (GB).

Colección e identificación

Los briófitos son muy fáciles de colectar, solo requieren bolsas de papel que se deben dejar abiertas para que el material se seque despues de colectado, además, no deben ser prensados y en lo posible se debe tratar de no mezclar especies. Recuerde que las especies más abundantes pueden no ser las más interesantes - muchas especies ocurren como individuos aislados. No es necesario colectar plantas con esporófitos!, cerca del 80–90 % de los musgos Ecuatorianos pueden ser determinados solo con gametófito. El equipo básico necesario para identificar musgos, incluye microscopio de disección y de luz. Es bueno primero al examinar los musgos, observar características generales, tales como la forma de crecimiento, patrones de ramificación, posición de las hojas (húmedas y secas), posición de los esporófitos (terminales o laterales sobre el gametófito), seta (lisa o papilosa), cápsulas (erectas o curvadas) etc., en el microscopio de disección. Luego se coloca una pequeña porción de la planta (dependiendo del tamaño) en agua hirbiendo, luego se extraen unas pocas hojas del tallo y se colocan sobre un portaobjetos y se coloca el cubreobjetos adicionando agua para llenar el área; usando un microscopio de luz, se puede observar mejor la forma de las hojas, los patrones celulares, etc. En general los musgos aquí descritos se pueden determinar por la forma y estructura de las hojas.

Es importante distinguir entre musgos y hepáticas foliosas las cuales a menudo son similares. La siguiente tabla muestra algunas diferencias entre estas.

Musgos	Hepáticas foliosas
Gametófito:	
Hojas generalmente con nervadura.	Hojas sin nervadura.
Hojas generalmente sin lóbulos.	Hojas lobuladas.
Células de la lámina isodiamétricas o alargadas.	Células de la lámina solo isodiamétricas.
Hojas dispuestas en 3-filas, generalmente en espiral.	Hojas dispuestas en 3-filas, dos laterales y una ventral.
Hojas generalmente mas largas que anchas, comunmente puntiagudas.	Hojas generalmente con igual ancho y largo, raramente puntiagudas.
Esporófito:	
Cuerpos de aceite ausentes.	Cuerpos de aceite presentes.
Opérculo y peristoma presentes en las capsulas.	Opérculo y peristoma ausentes.
Seta en su mayoría elongada, alambrina y persistente.	Seta a menudo corta, si se encuentra elongada, entonces débil y efímera.

Tratamiento taxonómico

Las especies aqui incluidas se basan en reportes de literatura o colecciones recientes encontradas en AAU, GB, NY y S. Referencias taxonómicas específicas son listadas en el tratamiento taxonómico de cada familia o género. Se proporcionan igualmente, claves en español para las familias, géneros y especies, que aparecen en letras itálicas. La nomenclatura generalmente sigue a Churchill (1989). Las ilustraciones están principalmente basadas en colecciones del amazonas.

Programas de investigación

Muchos proyectos son posibles y útiles con respecto a los musgos del amazonas. Las colecciones generales son siempre importantes ya que proporcionan información acerca de la distribución y podrían representar reporte de nuevas especies. Además, estudios de inventario en áreas de igual tamaño (p. e., parcelas de una hectárea) pueden ofrecer diverso tipo de información que podría ser de interés general. Por ejemplo, se podrían comparar dos o tres parcelas en bosques de tierra firme con otras tantas en bosques planos inundados; estas comparasiones podrían dirigirse i) al número de especies presentes, ii) estratificación y tipos de habitat, iii) formas de crecimiento, iv) biología reproductiva, p. e., reproducción sexual contra asexual, etc. Tal información podría proporcionar importantes contribuciones a nuestro conocimiento de los musgos en Ecuador y Neotrópicos en general. Estas son

---

Tabla 1. Formas de crecimiento de los musgos tropicales, tomada y modificada de Richards (1984), con ejemplos de musgos en la amazonía ecuatoriana.

A. Formas agrupadas. Tallos frondosos agregados.
  1. Cojines (tallos principales erectos y luego radiados para formar masas en forma de cúpula; cojines grandes = > 5 cm de diámetro, cojines prequeños = < 5 cm de diámetro); p. e., *Leucobryum*.
  2. Racimos (tallos erectos, ±paralelos).
    a. Racimos altos (> 2 cm de altura); p.e., *Holomitrium*, *Leucoloma*, algunas especies de *Octoblepharum*.
    b. Racimos erectos con ramas divergentes o rastreras; p.e. *Acroporium*, Orthotrichaceae.
    c. Racimos abiertos (< 2 cm de altura); p.e., *Fissidens*.
  3. Esteras (tallos primarios horizontalmente rastreros sobre el substrato, ramas laterales erectas o paralelas, usualmente formando esteras entretejidas).
    a. Esteras ásperas (tallos principales adheridos al substrato, con tallos laterales cortos y erectos); p.e., *Leucomium*, *Sematophyllum*.
    b. Esteras suaves (ramas en el mismo plano de los tallos principales); p.e., muchas Callicostaceae (incluyendo algunas especies de *Lepidopilum*), Hypnaceae, *Taxithelium*.
    c. Tejidas; p.e., *Thuidium*.
B. Formas solitarias. Tallos frondosos no cercanamente agregados.
  1. Musgos protonemales (protonema persistente, tallos frondosos menudos, dispersos); no presentes en el amazonas.
  2. Dendroides, no dorsiventrales; p.e., *Pireella*, *Porotrichum*, *Pseudocryphaea*.
  3. Formas de pluma (tallos secundarios ±pinnadamente ramificados, dorsiventrales); p.e., *Callicosta*, algunas especies de *Lepidopilum*, *Neckeriopsis*.
  4. Formas pendientes (tallos secundarios usualmente largos y pendulosos); p.e., Meteoriaceae, *Hildebrandtiella* y *Orthostichopsis*.

---

importantes preguntas que deberian ser respondidas por aquellos que residen en Ecuador.

3. KEY TO THE FAMILIES

1. Plants acrocarpous, forming tufts or cushions, primary stems erect; sporophytes mostly terminal at stem tip; peristome single, double or reduced; leaf costae single (appearing absent in Leucobryaceae)   **2.**
1. Plants pleurocarpous, primary and/or secondary stem procumbent, trailing, or pendent; sporophytes produced laterally on stems; peristome double or reduced; leaf costae single, double, short and forked or absent   **12.**
2. Leaves 2 or 4 ranked, not spirally arranged about stem, plants appearing flattened (complanate)   **3.**
2. Leaves 3-ranked or more, appearing spirally arranged, lamina not differentiated as above; peristome single or double   **4.**
3. Leaves 2-ranked, ±symmetric, not falcate, leaf lamina differentiated with a vaginant lamina base and extended dorsal and excurrent lamina; propagula not produced on distal stems   **Fissidentaceae (p. 72)**
3. Leaves appearing 4-ranked, falcate and asymmetric, lacking differentiated lamina as above; propagula usually present on distal stem tips   **Phyllodrepaniaceae (p. 109)**
4. Plants white or variously whitish-green or shinny red or purplish; in cross-section lamina cells differenitated with outer layer(s) of hyaline and empty cells (leucocysts) and an inner single layer of green cells (chlorocysts), costae obscure or appearing absent; peristome single   **Leucobryaceae (p. 89)**
4. Plants green or variously greenish-brown or golden yellow; lamina not differentiated with leucocysts and chlorocysts, costae distinct; peristome single or double   **5.**
5. Leaf margins doubly serrate; sporophyte at base of leafy gametophyte; peristome double   **Rhizogoniaceae (p. 125)**
5. Leaf margins singlely serrate (teeth not paired) or entire; sporophyte subterminal or terminal on leafy gametophyte; peristome double or single   **6.**
6. Marginal cells often linear (forming a weak to strong border), differentiated from inner lamina cells which are often isodiametric to hexagonal   **7.**
6. Marginal cells similar to inner lamina cells   **9.**
7. Median cells rhomboidal to broadly or narrowly hexagonal or fusiform, smooth throughout; peristome double   **Bryaceae (p. 34)**
7. Median cells mostly quadrate, smooth or papillose; peristome single   **8.**
8. Lower and basal cells sharply differenitated from distal cells; leaf base often sheathing   **Calymperaceae (p. 57)**
8. Lower and basal cells gradualy differentiated, becoming elongate rectangular (border absent); alar cells differentiated; leaf base not sheathing   **Dicranaceae (p. 67)**
9. Median cells papillose on distal cell angles; terminal stems often appearing fasiculate with several short branches; capsules subglobose; peristome double   **Bartramiaceae (p. 31)**

9. Median cells smooth or papillose, papillae over cell lumen;terminal stems not fasiculate; capsules ovoid tocylindrical; peristome single   **10.**
10. Leaves ovate- or oblong-subulate; basal cells often differentiated, cells enlarged   **Dicranaceae** (p. 67)
10. Leaves lanceolate to spatulate; basal cells not differentiated   **11.**
11. Leaves ligulate to ±spatulate; margins plane, cells smooth
                                               Splachnobryaceae (p. 133)
11. Leaves lanceolate, oblong-lanceolate or obovate; margins usually recurved; cells papillose or smooth   **Pottiaceae** (p. 111)
12. Costa absent or very short (1/5 lamina length or less) and forked   **13.**
12. Leaves costate, (1/4)1/3 or more lamina length, single or double   **22.**
13. Secondary stems often pendent; leaf base sub- to distinctly auriculate or not (*Hydropogon*)   **14.**
13. Secondary stems procumbent, spreading; leaf base not auriculate   **17.**
14. Secondary stem leaves most void of leaves or widely spaced on stem, distal branches densely foliate; leaf base not auriculate
                                               Hydropogonaceae (p. 78)
14. Secondary stems foliate, distal branches usually similarly foliate; leaf base often subauriculate to auriculate   **15.**
15. Leaves appearing 2-ranked, concave compressed
                                               Phyllogoniaceae (p. 109)
15. Leaves radially foliate about stem, not flattened, or if flattened then leaves differentiated between lateral and median   **16.**
16. Apex piliferous; leaves distinctly in 5-spirally arranged rows
                                               Pterobryaceae (p. 114)
16. Apex short acute, often recurved; leaves not distinctly arranged in 5 rows   **Meteoriaceae** (p. 95)
17. Alar cells differentiated, usually quadrate or inflated and oval   **18.**
17. Alar cells not differentiated   **20.**
18. Alar cells unequally differentiated on one side of lateral leaves; median cells smooth   **Stereophyllaceae** (p. 135)
18. Alar cells equally differentiated on both sides; median cells smooth or papillose   **19.**
19. Leaves often homomallous; alar cells usually inflated, mostly oval; plants forms mats or tufts   **Sematophyllaceae** (p. 125)
19. Leaves often falcate; alar cells quadrate or short rectangular, not inflated; plants forming mats   **Hypnaceae** (p. 79)
20. Leaf margins serrulate to near base, marginal teeth often bifid; clustered cylindrical propagula often beneath stems or on specialized branch tips
                                               Callicostaceae (p. 35)
20. Leaf margins smooth; propagula absent   **21.**
21. Leaves ovate, apices short acuminate; plants usually glossy green
                                               Hypnaceae (p. 79)
21. Leaves ovate-lanceolate to lanceolate, apices long acuminate; plants whitish-green   **Leucomiaceae** (p. 94)
22. Costae double, 1/4–4/5 lamina length, diverging or parallel from the base   **Callicostaceae** (p. 35)
22. Costa single, 1/2 or more lamina length   **23.**

23. Leaves bordered by linear cells, inner lamina cells hexagonal to isodiametric **24.**
23. Leaves not bordered, marginal cells ±similar to median **25.**
24. Plants course and stiff; secondary stems short and erect, marginal border confined to base; median cells isodiametric, thick-walled; plants dark green **Orthotrichiaceae** (p. 105)
24. Plants delicate and lax; secondary stems spreading to ±subascending; marginal border throughout; median cells hexagonal-rounded, ±thin-walled and lax; plants pale green to whitish-green **Daltoniaceae** (p. 65)
25. Leaves broadly oblong-ligulate, apex truncate; plants strongly complanate and perpendicular to substrate **Neckeraceae** (p. 103)
25. Leaves ovate, ovate-lanceolate or oblong-ligulate, apex acute, acuminate or obtuse; plants not or weakly complanate **26.**
26. Paraphyllia present on stems, papillose; leaf lamina cells uni- to pluripapillose, papillae often sharply pointed **Thuidiaceae** (p. 139)
26. Paraphyllia absent; leaf cells smooth, mamillose or papillose, papillae projecting at cell angles, or if over cell lumen then not sharply pointed **27.**
27. Leaves dimorphic, lateral leaves ovate-oblong, upperside leaves small, triangular-lanceolate; costae ±long excurrent **Racopilaceae** (p. 123)
27. Leaves ±similar; costae 2/3 lamina length or percurrent **28.**
28. Plants dendroid or frondose and usually stipitate, or pendent; alar region ±differentiated, cells often quadrate **29.**
28. Plants procumbent, not dendroid or frondose, secondary stems erect or spreading; alar region not distinct **31.**
29. Secondary stems ofen pendent with lamina cells smooth, or if plants spreading then lamina cells papillose, papillae over cell lumen **Meteoriaceae** (p. 95)
29. Secondary stems frondose or dendroid with lamina cells smooth or if papillose then projecting at cell angles, or if stems simple then papillae over cell lumen **30.**
30. Apex broadly acute to ±obtuse-acute, apical lamina cells short, rhombod to rhomboidal, 1–2(3): 1 **Thamnobryaceae** (p. 135)
30. Apex abruptly and ±narrowly acuminate or acute, apical lamina cells more elongate, 3–5: 1 **31.**
31. Apex narrowly to ±broadly acuminate; leaf base auriculate; microphyllous branches absent (branchs often flagellate) **Pterobryaceae** (p. 114)
31. Apex broadly acuminate or acute; leaf base not auriculate; microphyllous branches generally present **Leptodontaceae** (p. 87)
32. Leaves oblong-ligulate or -lanceolate; median cells isodiametric; margins entire **Orthotrichaceae** (p. 105)
32. Leaves broadly ovate to ovate-short lanceolate; median cells linear or oblong; margins serrulate or serrate **33.**
33. Leaf apices long acuminate, piliferous or acute-apiculate, median cells papillose, papillae over cell lumen or projecting at cell angles; margins distally serrate or weakly serrulate at base **Meteoriaceae** (p. 95)

33. Leaf apices acute, median cells smooth; margins serrate throughout,
particularly branches leaves   **Brachytheciaceae** (p. 32)

4. CLAVE PARA LAS FAMILIAS

1. Plantas acrocárpicas, formando racimos o cojines, tallos primarios erectos; la mayoría de los esporófitos terminales hacia el extremo del tallo; perístoma simple, doble o reducido; hojas con costa simple (ausentes en Leucobryaceae) **2.**
1. Plantas pleurocárpicas, tallo primario y/o secundario procumbente, rastrero, o pendiente; esporófitos producidos lateralmente sobre los tallos; perístoma doble o reducido; hojas con costa simple, doble, corta y dividida o ausente **12.**
2. Hojas dispuestas en 2–4 filas, no dispuestas en espiral sobre el tallo; plantas aplanadas (complanadas), perístoma simple **3.**
2. Hojas dispuestas en 3-filas o más, espiraladamente dispuestas sobre el tallo; lámina no diferenciada como en la anterior; perístoma simple o doble **4.**
3. Hojas dispuestas en 2-filas, ±simétricas, no falcadas, lámina diferenciada con una lámina basal conduplicada y una lámina dorsal extendida; propágulos no producidos sobre tallos distales **Fissidentaceae (p. 72)**
3. Hojas dispuestas en 4-filas, asimétricas falcadas y; lámina no diferenciada como en la anterior; propágulos usualmente presentes sobre los extremos distales de los tallos **Phyllodrepaniaceae (p. 109)**
4. Plantas blancas, verde-blancuzcas, rojo brillantes o purpuráceas; células de la lámina en sección transversal, diferenciadas con capa(s) externa de leucocistes y una única capa interna de clorocistes, costas obscuras o ausentes; perístoma simple **Leucobryaceae (p. 89)**
4. Plantas verdes, café-verduzcas o amarillo brillante; lámina no diferenciada, sin leucocistes y clorocistes, costas bien definidas; perístoma simple o doble **5.**
5. Márgenes de la hoja doblemente serrados; esporófito hacia la base del gametófito frondoso; perístoma doble **Rhizogoniaceae (p. 125)**
5. Márgenes de la hoja simplemente serrados (dientes no en pares) o enteros; esporófito terminal o subterminal sobre gametófito frondoso; perístoma doble o simple **6.**
6. Células marginales a menudo lineares, diferenciadas de las células al interior de la lámina, las cuales son a menudo isodiamétricas hasta hexagonales **7.**
6. Células marginales similares a aquellas células al interior de la lámina **9.**
7. Células medias romboidales hasta amplia o estrechamente hexagonales o fusiformes, completamente lisas; perístoma doble **Bryaceae (p. 34)**
7. Células medias en su mayoría cuadradas, lisas o papilosas; perístoma simple **8.**
8. Células basales e inferiores fuertemente diferenciadas a partir de las células distales; células alares no diferenciadas; base de la hoja a menudo envainada **Calymperaceae (p. 57)**
8. Células basales e inferiores gradualmente diferenciadas, llegando a ser rectangularmente elongadas; células alares diferenciadas; base de la hoja no envainada **Dicranaceae (p. 67)**

9. Células medias papilosas sobre los ángulos distales de las células; tallos terminales a menudo fasciculados con varias ramas cortas; cápsulas subglobosas; perístoma doble   **Bartramiaceae** (p. 31)
9. Células medias lisas o papilosas, papila sobre el lumen celular; tallos terminales no fasciculados; cápsulas ovoides hasta cilíndricas; perístoma simple   **10.**
10. Hojas ovadas u oblongo-subuladas; células basales a menudo diferenciadas, células agrandadas   **Dicranaceae** (p. 67)
10. Hojas lanceoladas a espatuladas; células basales no diferenciadas   **11.**
11. Hojas liguladas hasta ±espatuladas; márgenes planos, células lisas
                                                          Splachnobryaceae (p. 133)
11. Hojas lanceoladas, oblongo-lanceoladas u obovadas; márgenes recurvados; células papilosas o lisas   **Pottiaceae** (p. 111)
12. Costa ausente o muy corta (1/5 parte de la longitud de la lámina o menos) y dividida   **13.**
12. Hojas costadas, (1/4) 1/3 parte o más de la longitud de la lámina, simple o doble   **22.**
13. Tallos secundarios a menudo pendientes; base de la hoja sub hasta distintivamente auriculada o no (Hydropogon).   **14.**
13. Tallos secundarios procumbentes, postrados; base de la hoja no auriculada   **17.**
14. Tallo secundario en su mayoría desprovisto de hojas o ampliamente espaciadas sobre el tallo, ramas distales densamente foliadas; base de la hoja no auriculada   **Hydropogonaceae** (p. 78)
14. Tallo secundario foliado, usualmente las ramas distales similarmente foliadas; base de la hoja a menudo subauriculada hasta auriculada   **15.**
15. Hojas dispuestas en 2-filas, aplanadamente cóncavas
                                                          Phyllogoniaceae (p. 109)
15. Hojas radialmente foliadas con respecto al tallo, no aplanadas o si se encuentran aplanadas, entonces las hojas diferenciadas en laterales y medias   **16.**
16. Apice pilífero; hojas claramente dispuestas en 5-filas en espiral
                                                          Pterobryaceae (p. 114)
16. Apice cortamente agudo, a menudo recurvado; hojas no claramente dispuestas en 5-filas   **Meteoriaceae** (p. 95)
17. Células alares diferenciadas, usualmente cuadradas o infladas y ovaladas
                                                                                    18.
17. Células alares no diferenciadas   **20.**
18. Células alares desigualmente diferenciadas sobre uno los lados de las hojas laterales; células medias lisas   **Stereophyllaceae** (p. 135)
18. Células alares igualmente diferenciadas en ambos lados; células medias lisas o papilosas   **19.**
19. Hojas a menudo homomalas, células alares usualmente infladas, en su mayoría ovaladas ; plantas en forma de estera o racimo
                                                          Sematophyllaceae (p. 125)
19. Hojas a menudo falcadas; células alares cuadradas o cortamente rectangulares, no infladas; plantas formando esteras   **Hypnaceae** (p. 79)

20. Márgenes de las hojas serrulados hasta cerca a la base, dientes marginales a menudo bífidos; propágulos cilíndricos formando racimos, a menudo agrupados debajo de los tallos o sobre los extremos de ramas especializadas **Callicostaceae** (p. 35)
20. Márgenes de las hojas lisos; propágulos ausentes **21.**
21. Hojas ovadas; ápices cortamente acuminados; plantas usualmente verde-lustrosas **Hypnaceae** (p. 79)
21. Hojas ovado-lanceoladas a lanceoladas; ápices largamente acuminados; plantas verde-blancuzcas **Leucomiaceae** (p. 94)
22. Costa doble, 1/4–4/5 partes de la longitud de la lámina, divergentes o paralelas desde la base **Callicostaceae** (p. 35)
22. Costa simple, 1/2 o más de la longitud de la lámina **23.**
23. Hojas bordeadas por células lineares, células al interior de la lámina hexagonales a isodiamétricas **24.**
23. Hojas no bordeadas, células marginales ±similares a las medias **25.**
24. Plantas burdas y tiesas; tallos secundarios cortos y erectos; borde marginal solo hacia la base; células medias isodiamétricas, paredes engrosadas; plantas verde oscuras **Orthotrichaceae** (p. 105)
24. Plantas delicadas y laxas; tallos secundarios postrados hasta ±ascendentes; borde a lo largo de todo el márgen; células medias hexagonal-redondeadas, paredes ±delgadas y laxas; plantas verde-pálidas hasta verde-blancuzcas **Daltoniaceae** (p. 65)
25. Hojas ampliamente oblongo-liguladas, ápice truncado; plantas fuertemente complanadas y perpendiculares al substrato **Neckeriaceae** (p. 103)
25. Hojas ovadas, ovado-lanceoladas u oblongo-liguladas, ápice agudo, acuminado u obtuso; plantas ligeramente complanadas o no **26.**
26. Parafilos presentes sobre tallos, papilosos; células de la hoja uni-hasta pluripapilosas, papilas a menudo fuertemente puntiagudas **Thuidiaceae** (p. 139)
26. Parafilos ausentes; células de la hoja lisas, mamilosas o papilosas, papilas proyectandose hacia los ángulos de la célula, o si lo hacen sobre el lumen de la célula, entonces no fuertemente puntiagudas **27.**
27. Hojas dimórficas; hojas laterales ovado-oblongas, hojas del lado superior pequeñas, triangular-lanceoladas; costas ±largamente excurrentes **Racopilaceae** (p. 123)
27. Hojas ±similares; costas 2/3 partes de la longitud de la lámina o percurrentes **28.**
28. Plantas dendroides o frondosas y usualmente estipitadas, o pendientes; región alar ±diferenciada; células a menudo cuadradas **29.**
28. Plantas procumbentes, no dendroides o frondosas, ramas secundarias erectas o postradas; región alar no bien definida **32.**
29. Tallos secundarios pendientes; células de la lámina lisas, o si las plantas son postradas, entonces las células de la lámina papilosas, papilas sobre el lumen celular **Meteoriaceae** (p. 95)
29. Tallos secundarios frondosos o dendroides; células de la lámina lisas o si son papilosas, entonces estas se proyectan hacia los ángulos de la célula,

o si el tallo es simple, las papilas se encuentran sobre el lumen celular
                                                                          30.
30. Apice ampliamente agudo hasta ±obtuso-agudo, células apicales de la
    lámina cortas, romboidales, 2(3): 1   **Thamnobryaceae** (p. 135)
30. Apice abrupta y ±estrechamente acuminado o agudo, células apicales de
    la lámina más elongadas, 3–5: 1   **31.**
31. Apice estrecha hasta ampliamente acuminado; base de la hoja
    auriculada; ramas microfilas ausentes, ramas a menudo flageladas
                                                          Pterobryaceae (p. 114)
31. Apice ampliamente acuminado o agudo; base de la hoja no auriculada;
    ramas microfilas generalmente presentes   **Leptodontaceae** (p. 87)
32. Hojas oblongo-liguladas o lanceoladas; células medias isodiamétricas;
    márgenes enteros   **Orthotrichaceae** (p. 105)
32. Hojas ampliamente ovadas-cortamente lanceoladas; células medias
    lineares u oblongas; márgenes serrulados o serrados   **33.**
33. Apice de la hoja largamente acuminado, pilífero o agudo-espiculado;
    células medias papilosas, papilas sobre el lumen celular o proyectandose
    hacia los ángulos de las células; márgenes distalmente serrados o
    debilmente serrulados hacia la base   **Meteoriaceae** (p. 95)
33. Apice de la hoja agudo; células medias lisas; márgenes serrados a todo lo
    largo, particularmente en hojas ramificadas   **Brachytheciaceae** (p. 32)

5. TAXONOMIC TREATMENT

The species included in this florula are based on literature reports (Steere, 1948; Crum, 1957; Bartram, 1964; Robinson *et al.*, 1971) and recent collections by the present author and others found primarily at AAU, GB and NY. Additional species are expected for the Amazonas of Ecuador not treated in this present guide. If material from Amazonas does not match the descriptions, keys and illustations, further literature or assistance will be necessary. Floristic treatments consulted for this study and, in general, useful for Ecuador, for at least some of the common and widespread species are those of Bartram (1949), Florschütz (1964), Florschütz-de Waard (1986), Griffin (1979) and Mitten (1869). Specific taxonomic references are listed with the treatment of each family or genus.

Short descriptions are given for families and genera with keys only to species; variation described pertain only to taxa found in Amazonas. Characters given in the descriptions for families are generally not repeated under the genus. If only a single species is represented, the description follows the species name. In a very few cases it has been necessary to provide a description based on non-Amazonas collections or literature. The sequence of families, genera and species are arranged alphabetically. Information on distribution is provided for the known Amazonas provinces for each species based on literature and specimens examined primarily at AAU, but also at GB, NY and S. Not all specimens cited in the literature reports have been examined for this study, and thus for several species it has not been possible to provide verification of previous determinations or complete keys. Genera or species likely to be found in Ecuadorean Amazonas but for which there are presently no collections are denoted in the keys by an asterisk (*). Nomenclature generally follows that of Churchill (1989) unless otherwise stated. Synonyms are given only if they differ from those publications on Ecuador cited above, or if recently recognized in revisionary studies. Reference to previous reports based on the literature follow specimen citations.

Illustrations are primarily based on Amazonas collections but in some cases specimens from non-Amazonas Ecuador or Colombia have been used. It is hoped that the illustrations in themselves will be sufficient enough means to identify most collections.

The glossary, supplemented by illustrations, should be suffecent to use the present treatment. It is assumed that those using this

treatment have a basic background to general botanical
terminology.

BARTRAMIACEAE

Philonotis Brid.

Plants erect, solitary or more often forming short turfs; stems simple usually with terminal short fasciculate branches; leaves lanceolate, oblong-lanceolate or oblong, apex acuminate to obtuse, margins plane or reflexed or recurved, denticulate to serrate, costae single, strong, subpercurrent to long excurrent, median cells oblong-linear to rectangular, usually papillose at upper ends, basal cells rectangular, usually smooth; mostly dioicous; setae elongate, smooth; capsule inclined to suberect, urn subglobose, striate when dry, peristome double, exostome teeth 16, papillose, endostome basal membrane high, cilia present; opercula conic; calyptrae cucullate, naked and smooth; spores papillose.

Philonotis is commonly found in exposed or shaded places on wet soil and rock. A troublesome genus in need of revision in the Neotropics. Approximately 14 species have been reported from Ecuador.

Key to the Species
1. Costa ending several cells below apex; apex obtuse or bluntly acute
   P. gracillima
1. Costa percurrent to excurrent; apex acuminate **2.**
   2. Stem apices hooked or curved; leaves ±falcate; marginal teeth ±blunt; costa percurrent to short excurrent *P. uncinata*
   2. Stem apices erect; leaves usually straight; marginal teeth sharp; costa ±long excurrent *P. sphaerocarpa*

Clave para las Especies
1. Costa terminando varias células más abajo del ápice; ápice obtuso o despuntadamente agudo *P. gracillima*
1. Costa percurrente hasta excurrente; ápice acuminado **2.**
   2. Apices del tallo en forma de anzuelo o curvados; hojas ±falcadas; dientes marginales ±despuntados; costa percurrente hasta cortamente excurrente *P. uncinata*
   2. Apices del tallo erectos; hojas usualmente rectas; dientes marginales afilados; costa ±largamente excurrente
   P. sphaerocarpa

Philonotis gracillima Ångstr., Oefv. K. Vet. Ak. Foerh. 33(4): 17. 1876.

Napo: Añangu, ca. 75 km east of Coca, ca. 00°32'S, 76°23'W, 245–325 m, *Churchill & Sastre-De Jesús 13843* (NY; Churchill *et al.*, 1992).

Reported from the Galapagos Islands.

Philonotis sphaerocarpa (Hedw.) Brid., Bryol. Univ. 2: 25. 1827.
Napo: El Napo, 500 m, *Benoist 2680* (Thériot, 1936).
It is very possible that this collection actually represents *P. uncinata* which appears to be more common in Amazonas.

Philonotis uncinata (Schwaegr.) Brid., Bryol. Univ. 2: 221. 1827. (Fig. 4A–B).
Napo: Archidona, 00°54'S, 77°48'W, *Holm-Nielsen & Jeppesen 1023, 1056, 1057, 1058* (AAU; Robinson *et al.*, 1971). **Pastaza:** Montalvo, within military camp, 02°05'S, 76°58'W, ca. 250 m, *Løjtnant & Molau 13449* (AAU). Illustration based on *Løjtnant & Molau 13449*.

BRACHYTHECIACEAE

Oxyrrhynchium (B.S.G.) Warnst.

Oxyrrhynchium remotifolium (Grev.) Broth., Nat. Pfl. 1(3): 1154. 1909. (Fig. 4C-D). Synonym: *Rhynchostegium remotifolium* (Grev.) Spruce.
Plants medium sized, forming mats; primary stems creeping; secondary stems spreading to subascending, leaves ovate to broadly ovate, 1.2–1.5 mm long, ca. 1 mm wide, apex short acuminate to acute, margins plane distally, slightly recurved at base, serrate throughout, costae ca. 3/4 lamina length, median cells fusiform to narrowly oblong, smooth, apical cells shorter, rhomboidal to rhomboid, alar region not differentiated; synoicous; sporophyte lateral, setae elongate, scabrous; capsules horizontal, urn ovoid, peristome double, exostome teeth 16, striate below, papillose distally endostome basal membrane high, segments 16, cilia 1–3; opercula rostrate; calyptrae cucullate, naked.
Approximately seven species have been recorded for Ecuador. The generic limits between the genera *Eurhynchium, Oxyrrhynchium* and *Rhynchostegium* are not clearly defined and species listed under these various genera have moved back and forth. Recently Robinson (1987) segregated and placed one of the more common montane forest species of *Rhynchostegium, R. scariosum*, into a new genus, *Steerecleus*.
Pastaza: Río Bobonaza, 460 m, *Spruce 1416* (NY; Mitten, 1869). Illustration based on *Buck 4603* from Dominican Republic.
In Ecuador reported also from Chimborazo, Cotopaxi, Pichincha, and Tungurahua; at elevations from 460–2460 m.

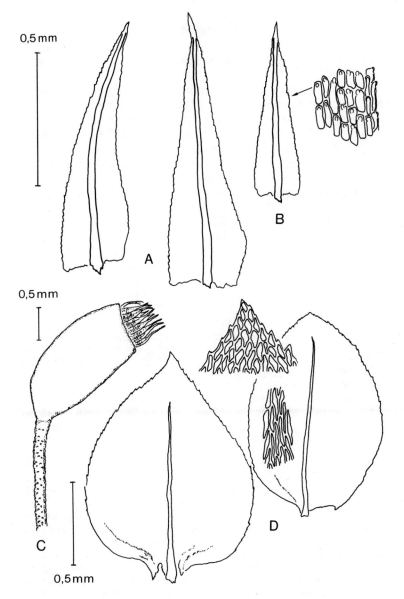

Figure 4. A-B. *Philonotis uncinata*, A. stem leaves, B. branch leaf; C–D. *Oxyrrhynchium remotifolium*, C. upper seta and capsule, D. stem leaves.

BRYACEAE

Bryum Hedw.

Plants erect, solitary or more commonly forming turfs; leaves lanceolate, oblong-lanceolate or ovate, apex acuminate to acute, margins plane to recurved below, entire to serrate distally, costae percurrent to short excurrent, median cells rhombodial to hexagonal, smooth, mostly thick-walled, marginal cells often forming a border of linear cells or border indistinct but cells progressively narrower and more elongate, basal cells rectangular; setae elongate, capsules horizontal to pendulous, often with a short neck, peristome double, exostome teeth 16, mostly papillose, endostome basal membrane high, segments 16, papillose, cilia 2–3, opercula conic; calyptrae cucullate, smooth.

Bryum is characteristically found on exposed or occasionally shaded place on soil or rocks. The genus is a large one for Ecuador with about 20 species, occurring from páramo to premontane forests, less frequent in the lowlands. References: Ochi (1980, 1981).

Key to the Species

1. Leaf margin ±similar to inner lamina cells, progressively narrower toward the margin; inner lamina cells narrowly rhomboidal to hexagonal   *B. apiculatum*
1. Leaf margin bordered by linear cells, inner lamina cells broadly rhomboidal to hexagonal   2.
    2. Leaf margins weakly bordered; leaves lanceolate, apex acuminate; capsules pendulous   *B. cornatum*
    2. Leaf margins ±strongly bordered; leaves oblong to somewhat obovate, apex acute; capsules suberect to horizontal   *B. limbatum*

Clave para las Especies

1. Márgen de la hoja ±similar a las células al interior de la lámina, progresivamente angostadas hacia el márgen; células al interior de la lámina estrechamente romboidales hasta hexagonales
       B. apiculatum
1. Márgen de la hoja bordeado por células lineares, células al interior de la lámina ampliamente romboidales hasta hexagonales   2.
    2. Márgen de la hoja debilmente bordeado; hojas lanceoladas, ápice acuminado; cápsulas pendulosas   *B. cornatum*
    2. Márgenes de la hoja ±fuertemente bordeados; hojas oblongas hasta algo obovadas, ápice agudo; cápsulas suberectas hasta horizontales   *B. limbatum*

Bryum apiculatum Schwaegr., Spec. Musc. Suppl. 1(2): 102. 72. 1816. (Fig. 5d). Synonyms: *Pohlia apiculata* (Schwaegr.) Crum & Anderson; *Pohlia cruegeri* (Hampe ex C. Müll.) Andr.

Pastaza: Montalvo, within military camp, 02°05'S, 76°58'W, ca. 250 m, *Løjtnant & Molau 13449* (AAU). Illustration based on *Løjtnant & Molau 13449.*

Bryum cornatum Schwaegr., Spec. Musc. Suppl. 1(2): 103. 71. 1816. (Fig. 5A-B).

Napo: Añangu, ca. 75 km east of Coca, ca. 00°32'S, 76°23'W, 245–325 m, *Churchill & Sastre-De Jesús 13842* (NY; Churchill *et al.*, 1992). Illustration based on *Weir 390*, Colombia (NY).

In Ecuador also reported from Guayas, Loja, Morona-Santiago; at elevations from 245–1000 m.

Bryum limbatum C. Müll., Syn. 2: 573. 1851. Synonyms: *Bryum maynense* Spruce in Mitt., *Bryum socorrense* (Hampe) Mitt.

Napo: 2 km W of Archidona, 00°54'S, 77°48'W, 600 m, *Holm-Nielsen & Jeppesen 1022* (AAU; Robinson *et al.*, 1971). **Pastaza:** Río Bobonaza, *Spruce 301, 304*; Río Pastaza, *Spruce 302* (NY; Mitten, 1869).

In Ecuador also reported from Morona-Santiago; at elevations from ca. 300–1100 m.

CALLICOSTACEAE

Plant primary stems creeping; secondary stems spreading, ascending, or erect and frondose, often complanate, leaves ovate, ovate-oblong or -lanceolate, symmetrical, or if leafy stems complanate, then lateral leaves often asymmetrical and median leaves symmetrical, margins elimbate or limbate, costae double, short or elongate, parallel or diverging, or absent (*Crossomitrium*), lamina cells isodiametric to linear, smooth or papillose; cylindrical propagula occasionally present on the leaves or in leaf axils; sporophyte lateral, setae short or elongate, smooth, papillose or spinose, capsules erect to pendulous, peristome double, exostome teeth 16, papillose or furrowed and striate, endostome basal membrane low, segments 16, cilia usually absent or reduced; opercula rostrate; calyptrae mitrate or mitrate-campanulate, smooth or hairy.

The family as circumscribed here is modified following Buck (1987a); traditionally the family Hookeriaceae contained *Leskeodon* and the genera presented below, and placed *Callicosta* (*Pilotrichum*) in its own family, the Pilotrichaceae.

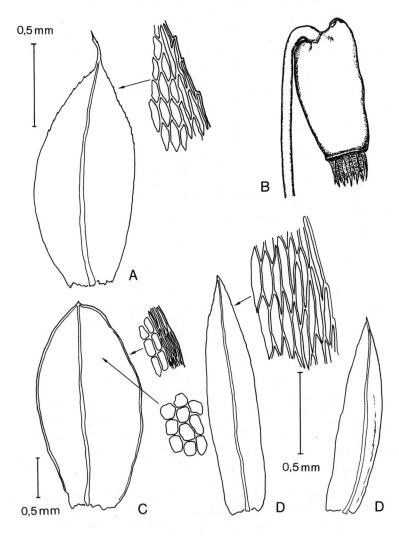

Figure 5. A-B. *Bryum cornatum*, A. stem leaf, B. capusle; C. *Dolotortula mniifolia*, C. stem leaf; D. *Bryum apiculatum*, D. stem leaves.

One of the largest families in the Neotropics and Ecuador, species of Callicostaceae commonly are epiphytic but also found on logs, and occasionally on rocks. Reference: Buck (1987a).

Key to the Genera

1. Leaves ecostate, appearing 4-ranked, leaf apex often pinched; propagula often present on underside of stems, clustered   *Crossomitrium*
1. Leaves costate, double; propagula present but not underside of stem or absent   **2.**
  2. Secondary stems erect, frondose, regularly pinnately branched, not complanate, leaves evenly distributed about stem   *Callicosta*
  2. Secondary stems spreading or ascending, irregularly and few branched, leaves variously complanate   **3.**
    3. Leaf median cells pluripapillose   **4.**
      4. Costae short, 1/4–1/3 lamina length; median cell papillae branched   *Hypnella*
      4. Costae long, 3/4 or more lamina length; median cell papillae simple, unbranched   *Pilotrichidium*
    3. Leaf median cells smooth or unipapillose   **5.**
      5. Leaf margins bordered by long linear cells in 1–3 rows, inner inner lamina cells broadly fusiform to hexagonal   **6.**
        6. Leaf cell walls lax; exostome furrowed, striate   *Cyclodictyon*
        6. Leaf cell walls firm, not lax; exostome appearing bordered, papillose   *Lepidopilum*
      5. Leaf margins cells similar to inner lamina cells   **7.**
        7. Leaves 3 times or more longer than wide; costa 2/3 lamina length and parallel; median cells distinctly porose   *Brymela*
        7. Leaves ±2 times longer than wide or less; costa diverging or parallel and merging distally; cells not or weakly porose   **8.**
          8. Median cells short, 1–2 times longer than wide, smooth or unipapillose; exostome furrowed, striate   *Callicostella*
          8. Median cells elongate, 4 × longer than wide, smooth; exostome appearing bordered, papillose   *Lepidopilum*

Clave para los Géneros

1. Hojas sin costa, dispuestas en 4-filas, ápice de la hoja a menudo contraido; propágulos a menudo presentes hacia el lado inferior de los tallos, agrupados   *Crossomitrium*
1. Hojas con costa doble; propágulos presentes pero no hacia el lado inferior del tallo o ausentes   **2.**

2. Tallos secundarios erectos, frondosos, regular y pinnadamente ramificados, no complanados, hojas uniformemente distribuidas en el tallo   *Callicosta*
2. Tallos secundarios postrados o ascendentes, poco e irregularmente ramificados, hojas variadamente complanadas   **3.**
    3. Células medias de la hoja pluripapilosas   **4.**
        4. Costas cortas, 1/4–1/3 parte de la longitud de la lámina; células medias con papilas ramificadas   *Hypnella*
        4. Costas largas, 3/4 partes o más de la longitud de la lámina; células medias con papilas simples, no ramificadas   *Pilotrichidium*
    3. Células medias de la hoja lisas o unipapilosas   **5.**
        5. Márgenes de la hoja bordeados por largas células lineares en 1–3 filas, células al interior de la lámina ampliamente fusiformes hasta hexagonales   **6.**
        6. Paredes celulares de la hoja laxas; exóstoma surcado, estriado   *Cyclodictyon*
        6. Paredes celulares de la hoja firmes, no laxas; exóstoma bordeado, papiloso   *Lepidopilum*
        5. Células marginales de la hoja similares a aquellas al interior de la lámina   **7.**
        7. Hojas 3 veces o más largas que anchas; costas 2/3 partes de la longitud de la lámina y paralelas; células medias distintivamente porosas   *Brymela*
        7. Hojas ±2 veces más largas que anchas o menos; costas divergentes o paralelas distalmente fusionadas; células debilmente porosas o no   **8.**
        8. Células medias cortas, 1–2 veces más largas que anchas, lisas o unipapilosas; exóstoma surcado, estriado   *Callicostella*
        8. Células medias elongadas, 4 veces más largas que anchas, lisas; exóstoma bordeado, papiloso   *Lepidopilum*

Brymela Crosby & Allen

Brymela acuminata (Mitt.) Buck, Brittonia 39: 217. 1987. (Fig. 6A-B).
Synonym: *Hookeriopsis acuminata* (Mitt.) Jaeg.
Plants rather large; stems and branches spreading; median and lateral leaves differentiated, lateral leaves undulate-crispate, oblong to ovate-oblong, 1.8–2.2 mm long, 0.6–0.8 mm wide, apex short acuminate, base subauriculate, margins plane to ±recurved, distal 2/3 serrulate, costae double, strong, ±parallel, ending below apex, costa apex ending in a sharp spine, median cells oblong-rectangular, ±porose, basal incertion cells golden-brown. Sporophytes not observed.

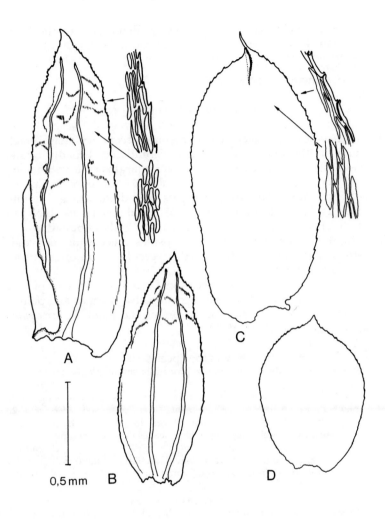

Figure 6. A-B. *Brymela acuminata*, A. lateral stem leaf, B. branch leaf; C-D. *Crossomitrium patrisiae*, C. lateral leaf, D. median leaf.

Brymela contains three species in Ecuador, mostly from premontane to low montane forests. Reference: Buck (1987a).

Pastaza: Río Bobonaza, 460 m, *Spruce 684* (NY; Steere, 1948). Illustration based on *Holm-Nielsen & Jeppesen 586* (AAU) from Ecuador.

At elevations from 460–1000 m.

Callicosta C. Müll.

Plants frondose; primary stems creeping secondary stems erect, pinnate or bipinnate, leaves not differentiated, broadly ovate, concave, short acute, weakly auriculate, margins recurved below, serrate to near base, costae strong, double, 3/4 or more the lamina length, ±parallel to slightly converging, weakly to strongly serrate on back, ending in a short spine, median and distal cells oblong to oval, thick-walled, alar region ±distinct, cells subquadrate to rectangular; setae short, smooth; capsules erect, short obovoid, exostome papillose, not furrowed; opercula rostrate; calyptrae campanulate, hairy.

The genus contains four species in Ecuador, in shaded places, commonly on trees and treelets, occasionally on lianas and palms. *Callicosta* was previously known under the name *Pilotrichum*. References: Crosby (1969, 1978).

Key to the Species

1. Costa back weakly serrate distally, propagula absent; plants synoicous
    C. bipinnata
1. Costa back strongly serrate to near base, propagula present along costa; plants dioicous   **2.**
    2. Subapical cells of branch leaves longer and thicker than surrounded cells; secondary stem leaves 0.8–1.2 mm long
        C. fendleri
    2. Subapical cells of branch leaves similar to surrounding cells; secondary stem leaves 0.6–0.7 mm long   *C. armata*

Clave para las Especies

1. Costa leve y serrada distalmente sobre el lado abaxial, propágulos ausentes; plantas sinoicas   *C. bipinnata*
1. Costa fuertemente serrada hasta muy cerca a la base sobre el lado abaxial, propágulos presentes a lo largo de la costa; plantas dioicas   **2.**
    2. Células subapicales de las hojas ramificadas más largas y gruesas que las células circundantes; hojas de tallos secundarios 0.8–1.2 mm de longitud   *C. fendleri*
    2. Células subapicales de las hojas ramificadas similares a las células circundantes; hojas de tallos secundarios 0.6–0.7 mm de longitud
        C. armata

Callicosta armata (Broth.) Crosby, Bryologist 81: 436. 1978.
Synonym: *Pilotrichum armatum* Broth.
Napo: El Napo, 500 m, *Benoist 4663* (Thériot, 1936).
The correct identity of this collection is still questionable, Crosby (1969) did not find the Benoist collection in Thériot's herbarium (PC). Given the close similarity with *C. fendleri*, the Benoist collection could be this taxon which aparently is more common in Amazonas.
In Ecuador known from Morona-Santiago and ?Mt. Guayrapurina; at elevations to 2100 m.

Callicosta bipinnata (Schwaegr.) C. Müll., Linnaea 21: 189. 1848.
(Fig. 7A-C). Synonym: *Pilotrichum bipinnata* (Schwaegr.) Mitt.
Morona-Santiago: Yurupaz, 600 m, *Harling 2262b* (Crum, 1957). Napo: Shinguipino, between Río Napo and Tena, ca. 460 m, *Grubb et al.* 2920a, 2925b (Bartram, 1964); Las Sachas, on road 40 km Coca–Lago Agrio, ca. 250 m, *Fransén 60* (AAU, GB); Añangu, ca. 75 km east of Coca, ca. 00°32'S, 76°23'W, 245–325 m, *Churchill & Sastre-De Jesús 13873* (NY; Churchill et al., 1992); near Tena, 6 km along Río Pano, 00°58'S, 77°52'W, 600 m, *Holm-Nielsen & Jeppesen 683*, (AAU, GB; Robinson et al., 1971). Pastaza: Río Bobonaza, 365 m, *Spruce 692* (NY; Mitten, 1869); Curaray, ridge NE of Destacamento, 01°21'S, 76°56'W, 250 m, *Holm-Nielsen et al. 21960* (AAU); Curaray, SE of airstrip, 01°22'S, 76°57'W, 250 m, *Holm-Nielsen et al. 22279* (AAU); Curaray, northern bank 2 km W of school, 01°22'S, 76°58'W, 250 m, *Holm-Nielsen et al. 21834* (AAU); Curaray, Valle de la Muerte, 01°25'S, 76°52'W, 240 m, *Holm-Nielsen et al. 22401, 22415, 22494* (AAU); Lorocachi, near military camp, 01°38'S, 75°58'W, 200 m, *Brandbyge & Asanza C. 30843, 31060* (AAU), *Jaramillo et al. 31146a, 31392* (AAU). Sucumbíos: Reserva Faunística Cuyabeno, near Laguna Grande, 00°00', 76°12'W, 265 m, *Balslev 84917* (AAU); *Heikkienen RH 1990–38* (NY). Illustration based on *Balslev 84917*.
In Ecuador reported from Chimborazo, Pichincha and Tungurahua; at elevations from 245–2770 m. Apparently the more common species of *Callicosta* in Amazonas.

Callicosta fendleri (C. Müll.) Crosby Bryologist 81: 436. 1978. (Fig. 7D-E). Synonym: *Pilotrichum fendleri* C. Müll.
Napo: Río Aguarico, E of the mouth of Río Cuyabeno, 0°16'S, 75°54'W, 200 m, *Holm-Nielsen et al. 21497* (AAU). Añangu, ca. 75 km east of Coca, ca. 00°32'S, 76°23'W, 245–325 m, *Churchill & Sastre-De Jesús 13836* (AAU, NY; Churchill et al., 1992). Sucumbíos: On Ecuadorian side of Río Putumayo, ca. 6 km downstream from Batallón de Selva N° 55, 00°05'N, 75°52'W, ca. 200 m, *Andrade 33170* (AAU). Illustration based on *Churchill & Sastre-De Jesús 13836*.
In Ecuador known from Pichincha.

Callicostella (C. Müll.) Mitt.

Plants forming loose to dense mats; stems spreading, leaves crispate when dry, complanate, lateral and median leaves differentiated, lateral leaves asymmetric, ovate to ovate-oblong,

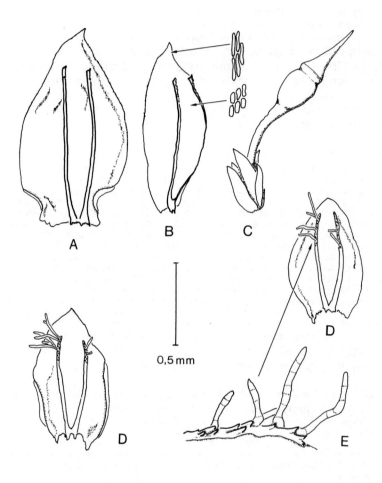

Figure 7. A-C. *Callicosta bipinnata*, A. stem leaf, B. branch leaf, C. sporophyte; D-E. *Callicosta fendleri*, D. stem leaves, E. propagulae along distal costa back.

short acuminate to nearly truncate-apiculate, margins serrate to dentate distally, costae double, 2/3–4/5 lamina length, usually strong, diverging to ±merging distally, smooth to serrate on back, median cells hexagonal to isodiametric, thick-walled, smooth to papillose (projecting angles and/or over cell lumen); setae elongate, smooth to papillose; capsules inclined to subpendent, ovoid, exostome striate below, papillose above, furrowed; opercula rostrate; calyptrae mitrate, lacerate at base.

Callicostella is most common in the wet lowland and premontane forests, often on decaying logs. Eight species are recorded for Ecuador.

Key to the Species

1. Apex acuminate; costae not strong; plants usually aquatic *C. rivulare*
1. Apex truncate or obtuse-rounded and apiculate; costae strong; plants usually terrestrial **2.**
    2. Median and distal cells distinctly papillose; costae ±parallel, or weakly merging *C. pallida*
    2. Median cells smooth, apical cells smooth or faintly papillose; costae ±merging distally *C. merkelii*

Clave para las Especies

1. Apice acuminado; costas débiles; plantas usualmente acuáticas *C. rivulare*
1. Apice truncado u obtuso-redondeado y apiculado; costas fuertes; plantas usualmente terrestres **2.**
    2. Células medias y distales distintivamente papilosas; costas ±paralelas o levemente fusionadas *C. pallida*
    2. Células medias lisas, células apicales lisas o debílmente papilosas; costas ±fusionadas distalmente *C. merkelii*

Callicostella merkelii (Hornsch.) Jaeg., Ber. St. Gall. Naturw. Ges. 1875–76: 351. 1877. (Fig. 8B). Synonym: *Schizomitrium merkelii* (Hornsch.) J. Florsch.

Pastaza Río Bobonaza, 460 m, *Spruce 634* (Mitten, 1869). **Sucumbíos:** Reserva Faunística Cuyabeno, near Laguna Grande, 00°00', 76°12'W, 265 m, *Balslev 84924* (AAU). Illustration based on *Balslev 84924.*

The report of *Callicostella saxatilis* (Mitt.) Jaeg. from Amazonas was based on a collection made by Spruce (no. 634). This collection, *Spruce 634*, and a further collection from Brazil, *Spruce 636*, are syntypes of *C. saxatilis* (Mitten, 1869). These are two different taxa, at least based on types examined from BM; *Spruce 636* matches the features given in the keys by Mitten (1869), however *Spruce 634* appears to match *Callicostella merkelii* described and illustrated by Florschütz-de Waard (1986), and that name is used in the present treatment.

Callicostella pallida (Hornsch.) Ångstr., Oefv. K. Vet. Ak. Foerh. 33: 27. 1876. (Fig. 8C-E). Synonyms: *Callicostella aspera* (Mitt.) Jaeg., *Schizomitrium pallidum* (Hornsch.) Crum & Anderson

Morona-Santiago: Patuca, 600 m, *Harling 2275 p.p.*, *2277* (S), *2280b p.p.* (Crum, 1957). **Napo:** Canton Napo, trail from Tena to Napo, 400 m, *Mexía 7170a* (GB); Río Aguarico, E of the mouth of Río Cuyabeno, 0°16'S, 75°54'W, 200 m, *Holm-Nielsen et al. 21518* (AAU); Añangu, ca. 75 km east of Coca, ca. 00°32'S, 76°23'W, 245–325 m, *Churchill & Sastre-De Jesús 13814, 13862* (AAU, NY; Churchill *et al.*, 1992). **Pastaza** Río Bobonaza, 460 m, *Spruce 626* (NY; Mitten, 1869); Finca Valle de Muerte on Río Curaray, ca. 1 hour downstream from Curaray, ca. 200 m, *Andersson 901c* (GB); Curaray, 2 km W of school, 1°22'S, 76°58'W, 250 m, *Holm-Nielsen et al. 21902* (AAU). **Sucumbíos:** Reserva Faunística Cuyabeno, near Laguna Grande, 00°00', 76°12'W, 265 m, *Balslev 84921* (AAU); *Heikkinen RH-1990–298* (NY). Illustration based on *Andersson 901c.*

In Ecuador also reported from Guayas and Pichincha; at elevations from 245–1000 m.

Callicostella rivularis (Mitt.) Jaeg., Ber. s. Gall. Naturw. Ges. 1875–76: 355. 1877. Synonym: *Schizomitrium rivulare* (Mitt.) Crum

Pastaza: Río Bobonaza, 460 m, *Spruce 668* (NY; Mitten, 1869).

In Ecuador reported from Morona-Santiago; at elevations from 460–1000 m.

Crossomitrium C. Müll.

Plants glossy green or golden, forming thin mats; secondary stems spreading, leaves complanate, when dry crispate or not, lateral and median leaves differentiated, lateral leaves oblong-obovate to oblong-lanceolate, or orbicular to ovate-oval, short acuminate or acute, apex usually pinched or not, margins serrate to base, teeth mostly bifid, ecostate, median cells short to long fusiform, smooth, median leaves smaller, obovate to suborbicular; propagula usually present, clustered beneath or on specialized branches, cylindrical; dioicous; setae elongate, distally papillose; capsules suberect, ovoid, exostome papillose, endostome cilia absent; opercula short conic-rostrate; calyptrae mitrate, divided at base.

Crossomitrium is one of the most common epiphyllous mosses in Amazonas, though it is also commonly found on trunks and branches of treelets and herbs, occasionally on logs. Four species are recorded for Ecuador. Reference: Allen (1990).

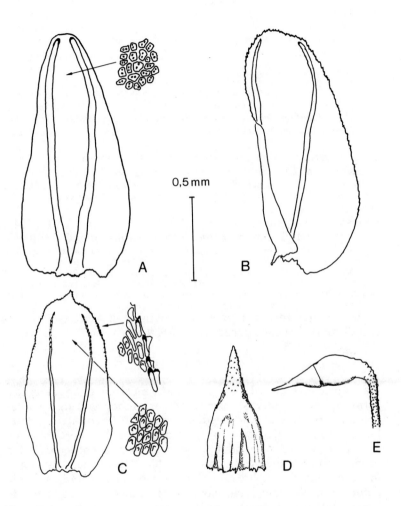

Figure 8. A. *Pilotrichidium callicostatum*, A. stem leaf; B. *Callicostella merkelii*, B. stem leaf; C-E. *Callicostella pallida*, C. stem leaf, D. calyptra, E. capsule.

Key to the Species

1. Lateral leaves orbicular to ovate-oval, when dry usually flat and close to substrate   *C. epiphyllum*
1. Lateral leaves oblong-short lanceolate, when dry often arching, ±crispate   *C. patrisiae*

Clave para las Especies

1. Hojas laterales orbiculares hasta ovado-oval, usualmente planas y próximas al substrato cuando están secas   *C. epiphyllum*
1. Hojas laterales oblongo-cortamente lanceoladas, a menudo arqueadas, ±crispadas cuando están secas   *C. patrisiae*

Crossomitrium epiphyllum (Mitt.) C. Müll., Linnaea 38: 613. 1874. Synonym: *Crossomitrium orbiculatum* C. Müll.

Napo: Añangu, 70 km E of Coca on the Río Napo, 300 m, *Brako 4619b* (NY); ca. 6.7 km S of Tena on road to Puyo, 01°02'S, 77°48'W, *Churchill & Sastre-De Jesús 13660* (NY).

Crossomitrium patrisiae (Brid.) C. Müll., Linnaea 38: 612. 1847. (Fig. 6C-D). Synonym: *Crossomitrium spruceanum* C. Müll.

Napo: Shinguipino, between Río Napo and Tena, ca. 460 m, *Grubb et al. 2924a* (Bartram, 1964); Añangu, ca. 75 km east of Coca, ca. 00°32'S, 76°23'W, 245–325 m, *Churchill & Sastre-De Jesús 13808, 13829, 13830, 13863* (AAU, NY; Churchill et al., 1992). **Pastaza:** Río Bobonaza, 365 m, *Spruce 790* (NY; Mitten, 1869). **Sucumbíos:** Reserva Faunística Cuyabeno, near Laguna Grande, 00°00', 76°12'W, 265 m, *Balslev 84919* (AAU); *Heikkinen RH-1990–25* (NY). Illustration based on *Churchill & Sastre-De Jesús 13829*.

In Ecuador reported from the provinces of Morona-Santiago and Pichincha; at elevations from 245–950 m.

Cyclodictyon Mitt.

Plants pale green to whitish-green, medium sized, often rather lax, forming mats; stems spreading or subascending, leaves complanate, lateral and median leaves differentiated, lateral leaves asymmetric, ovate to ovate-oblong, apex short to ±long acuminate, margins smooth to more often serrate, bordered by 1–5 or more rows of liner cells, costae double, 2/3–3/4 lamina length, median cells large, hexagonal to broadly rhomboid, smooth; autoicous or dioicous; setae elongate, smooth; capsules suberect to horizontal, urn ovoid to ovoid-cylindrical, ±curved, exostome furrowed, striate below, endostome papillose; opercula rostrate; calyptrae mitrate, smooth, slightly lacinate at base.

Cyclodictyon is a rather large and highly variable genus in Ecuador and the Neotropics. Although 18 species are recorded from Ecuador, this number may be reduced to half after careful study of the genus in the Neotropics. The features used by Mitten (1869) to distinguish our species, excluding *C. roridum* which exhibits a greater number of border cells than the remaining three species, are not particularly well defined given the variability in the group: *C. bombonasica* is said to have lateral leaves oblong-ligulate, apex short acuminate, and perichaetial leaves ovate-acuminate; *C. albicans* the leaves are oblong, apex acuminate, and perichaetial leaves subulate; and finally, *C. aeruginosa* the leaves are ovate-oblong, apex acuminate and perichaetial leaves subulate and serrate. Three of the species are treated here.

Key to the Species

1. Marginal border (4)6 cells wide   *C. roridum*
1. Marginal border 2(3) cells wide   **2.**
    2. Lateral leaves oblong-ligulate, 2.2–2.4 mm long; apices ±abruptly short acuminate   *C. bombonasicum*
    2. Lateral leaves ovate- to oblong-lanceolate, (1)1.5–1.8 mm long; apices gradually to ±abruptly acuminate   *C. albicans*

Clave para las Especies

1. Borde marginal de 4 (6) células de ancho   *C. roridum*
1. Borde marginal de 2 (3) células de ancho   **2.**
    2. Hojas laterales oblongo-liguladas, 2.2–2.4 mm de longitud; ápice ±abruptamente corto acuminado   *C. bombonasicum*
    2. Hojas laterales ovado- hasta oblongo-lanceoladas, (1)1.5–1.8 mm de longitud; ápice gradualmente hasta ±abruptamente acuminado   *C. albicans*

Cyclodictyon aeruginosum (Mitt.) O. Kuntze, Rev. Gen. Pl. 2: 835. 1891.

Pastaza: Río Bobonaza, 460 m, *Spruce 601, 605* (NY; Mitten, 1869).

In Ecuador reported also from Tungurahua; at elevations from 460–2460 m.

Cyclodictyon albicans (Hedw.) O. Kuntze, Rev. Gen. Pl. 2: 835. 1891. (Fig. 9A-B).

Napo: Reserva del Batallon de Selva No. 55, 0°05'N, 75°52'W, ca. 200 m, *Andrade 33093* (AAU); Añangu, ca. 75 km east of Coca, ca. 00°32'S, 76°23'W, 245–325 m, *Brako 5165* (NY). Illustration based on *Churchill et al. 15237* (NY) from Colombia.

In Ecuador reported from Chimborazo, Tungurahua and also the Galapagos Islands; at elevations from 245–2310 m.

Cyclodictyon bombonasicum (Mitt.) O. Kuntze, Rev. Gen. Pl. 2: 835. 1891.

Pastaza: Río Bobonaza, 460 m, *Spruce 598* (NY; Mitten, 1869).

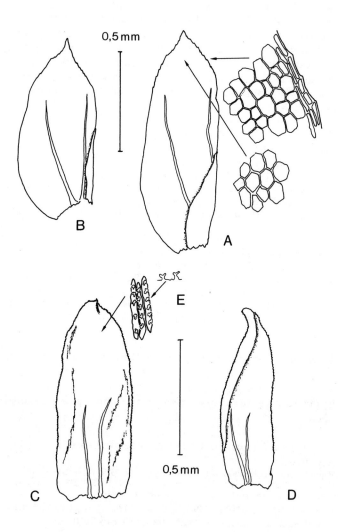

Figure 9. A-B. *Cyclodictyon albicans*, A. lateral stem leaf, B. median leaf; C-E. *Hypnella pallescens*, C. stem leaf, D. branch leaf, E. side view of papillae.

Cyclodictyon roridum (Hampe) O. Kuntze, Rev. Gen. Pl. 2: 835. 1891. Synonym: *Cyclodictyon riparium* (Mitt.) Broth.

Pastaza: Río Bobonaza, 365 m, *Spruce 593* (NY; Mitten, 1869).

Hypnella (C. Müll.) Jaeg.

Hypnella pallescens (Hook.) Jaeg., Ber. S. Gall. Naturw. Ges. 1875–1876: 365. 1877. (Fig. 9C-E).

Plants somewhat small, often glaucous green or whitish-green, forming rather dense mats; secondary stems spreading, loosely complanate, leaves broadly ovate to ovate-oblong, 0.6–1 mm long, 0.3–0.4 mm wide, concave, apex round-acute, margins appearing serrulate by marginal papillae, costae double and ±short, ca. 1/4–1/3 lamina length, median cells oblong-linear, pluripapillose, seriate, papillae usually branched; autoicous or synoicous; setae elongate, 2.2–2.8 cm long, distally roughened, capsules subpendulous, peristome double, exostome cross-striate, furrowed, endostome basal membrane ±high, cilia absent; opercula rostrate; calyptrae mitrate, ca. 2.2 mm long, split at base.

Three species occur in Ecuador, only *Hypnella pallescens* is likely to be found at lower elevations. References: Allen (1986), Crosby *et al.* (1985).

Pastaza: Río Bobonaza, 460 m, *Spruce 620* (NY; Mitten, 1869). Illustration based on *Spruce 615* (AAU) from Brazil.

Lepidopilum (Brid.) Brid.

Plants small to rather large, forming loose to dense mats or tufts, with secondary stems spreading or more often erect and ±perpendicular to substrate, radially foliate to more often complanate, lateral and median leaves differentiated, lateral leaves often asymmetric, ovate-lanceolate, obovate to oblong or oblong-lanceolate, some falcate and often sigmoid-shaped, acuminate to obtuse, margins elimbate to limbate, mostly serrate to serrulate distally, costae double, 1/5–3/4 lamina length, median cells linear or fusiform to hexagonal, marginal cells often forming a border of linear cells, median leaves usually smaller, symmetric; propagula often present in leaf axils, cylindrical; setae papillose to more often spinose distally or throughout, capsules mostly suberect, urn ovoid or ovoid-cylindrical, exostome teeth appearing border, endostome basal membrane low, cilia absent; operculum rostrate, calyptrae mitrate-campanulate, plicate, rarely smooth, entire or divided at base naked to more often hairy.

Lepidopilum is a rather large genus in Ecuador with about 30 species, commonly from wet lowland to high montane forests, typically epiphytic on trees, treelets, shrubs and lianas, occasionally on logs and with the few aquatic species, on rocks. References: Churchill (1988b, 1992).

Key to the Species

1. Lateral leaves limbate, 2–8 rows of linear cells along margin, inner lamina cells larger, fusiform to hexagonal **2.**
   2. Leaves ovate-lanceolate, apex long acuminate, costae 1/2 lamina length; setae short, 2–3 mm long, papillose distally
      L. polytrichoides
   2. Leaves ovate-oblong to oblong-lanceolate or obovate, apex obtuse to abruptly acuminate, apiculate or acute, costae 1/2–2/3 lamina length; setae short to elongate, spinose to papillose **3.**
      3. Leaves long oblong-obovate to ovate-oblong, 3–4 times longer than wide **L. arcuatum**
      3. Leaves short ovate-oblong to obovate, 2 times longer than wide **4.**
         4. Leaf border of 1–3 rows of linear cells, often inconspicuous, costae ca. 1/2 lamina length **5.**
            5. Leaves obovate, primary stems mostly tomentose; setae 2.5–3 mm long, smooth or distally roughened **L. surinamense**
            5. Leaves ovate, broadest near middle, primary stems not distinctly tomentose; setae 4.5–5 mm long, papillose throughout **L. curvifolium**
         4. Leaf border 4 rows or more of linear cells, costae 2/3–3/4 lamina length, apex apiculate; setae 8–12 mm long
            L. tortifolium
1. Lateral leaves elimbate or inner lamina cells progressively small toward margin **6.**
   6. Margins smooth throughout, not serrate **L. radicale**
   6. Margins serrulate or serrate, not forming a distinct border **7.**
      7. Lateral leaves symmetric or weakly asymmetric, ovate-oblong to ligulate, apex obtuse to acute; mat forming **8.**
         8. Lateral leaves small to medium sized, 1.4–2.5(3) mm long, 0.5–1 mm wide; epiphytic or lignicolous **L. affine**
         8. Lateral leaves large, 3.5–4 mm long, 1–1.3 mm wide; aquatic or semiaquatic, usually saxicolous
            L. pallido-nitens
      7. Lateral leaves strongly asymmetric, oblong-lanceolate, falcate, often S-shaped, apex short acuminate; plants often forming tufts **9.**

9. Setae 2–3.5 mm long; leaves often distantly spaced, stems often visible, deciduous leaves absent on terminal branches; leaf apex long acuminate
    L. brevipes
9. Setae 4 mm or longer; leaves usually crowded, stem usually not visible, deciduous leaves often present on terminal branches; leaf apex short acuminate
    L. scabrisetum

Clave para las Especies

1. Hojas laterales bordeadas por 1–8 filas de células lineares, células al interior de la lámina grandes, fusiformes hasta hexagonales **2.**
    2. Hojas ovado-lanceoladas, ápice largamente acuminado; costa 1/2 de la longitud de la lámina; setas cortas, 2–3 mm de longitud, distalmente papilosas **L. polytrichoides**
    2. Hojas ovado-oblongas hasta oblongo-lanceoladas u obovadas, ápice obtuso hasta abruptamente acuminado, apiculado o agudo; costa 1/2–2/3 partes de la longitud de la lámina; setas cortas a elongadas, espinosas hasta papilosas **3.**
        3. Hojas largamente oblongo-obovadas hasta ovado-oblongas, 3–4 veces más largas que anchas **L. arcuatum**
        3. Hojas cortamente ovado-oblongas hasta obovadas, 2 veces más largas que anchas **4.**
            4. Hojas bordeadas por 1–3 filas de células lineares, a menudo inconspicuas, costas ca. 1/2 de la longitud de la lámina **5.**
                5. Hojas obovadas, tallos primarios en su mayoría tomentosos; setas 2.5–3 mm de longitud, lisas o ásperas distalmente **L. surinamense**
                5. Hojas ovadas, ampliadas hacia la parte media, tallos primarios no distintivamente tomentosos; setas 4.5–5 mm de longitud, papilosas en toda su longitud **L. curvifolium**
            4. Hojas bordeadas por 4-filas o más de células lineares, costas 2/3–3/4 partes de la longitud de la lámina, ápice apiculado, setas de 8–12 mm de longitud **L. tortifolium**
1. Hojas laterales no bordeadas o las células al interior de la lámina progresivamente pequeñas hacia el márgen **6.**
    6. Márgenes lisos en toda su longitud, no serrados **L. radicale**
    6. Márgenes serrulados o serrados, sin formar un borde distintivo **7.**
        7. Hojas laterales simétricas o levemente asimétricas, ovado-oblongas hasta liguladas, ápice obtuso hasta agudo; plantas formando esteras **8.**
            8. Hojas laterales pequeñas hasta de talla media, 1.4–2.5 (3) mm de longitud, 0.5–1 mm de ancho; epífitas o lignícolas **L. affine**

8. Hojas laterales grandes, 3.5–4 mm de longitud, 1–1.3 mm de ancho; acuáticas o semiacuáticas, usualmente saxícolas **L. pallido-nitens**
7. Hojas laterales fuertemente asimétricas, oblongo-lanceoladas, falcadas, a menudo en forma de S , ápice cortamente acuminado, plantas a menudo formando racimos **9.**
9. Setas de 2–3.5 mm de longitud; hojas a menudo bien espaciadas, tallos a menudo visibles; hojas deciduas ausentes sobre las ramas terminales; ápice de las hojas largamente acuminado **L. brevipes**
9. Setas de 4 mm de longitud o más; hojas usualmente cercanas entre sí, tallos usualmente no visibles; hojas deciduas a menudo presentes sobre las ramas terminales; ápice de las hojas cortamente acuminado **L. scabrisetum**

Lepidopilum affine C. Müll., Linnaea 21: 192. 1848. (Fig. 10A-D). Synonyms: *Lepidopilum allionii* Broth., *Lepidopilum antisanense* Bartr., *Lepidopilum mittenii* C. Müll., *Lepidopilum pumilum* Mitt.

Morona-Santiago: Yurupaz, 600 m, *Harling 2241a,b* (S). **Napo:** Shinguipino, between Río Napo and Tena, ca. 460 m, *Grubb et al. 2925* (BM, type of *L. antisanense*), *2925a* (Bartram, 1964); SW of Nuevo Rocafuerte, along Río Braga, 200–230 m, *Jaramillo & Coello 4608 p.p.* (AAU; with *Thuidium campanulatum*); Añangu, ca. 75 km east of Coca, 00°32'S, 76°23'W, 245–325 m, *Churchill & Sastre-De Jesús 13781, 13786, 13791-b, 13793-d. 13798, 13806, 13831*(AAU, NY; Churchill et al., 1992). **Pastaza:** Río Bobonaza, 365 m, *Spruce 749* (NY; Mitten, 1869); Namoyacu at Río Curaray, 230 m, 01°27'S, 76°47'W, *Holm-Nielsen et al. 22286* (AAU); Montalvo, near military camp, 2°05'S, 76°58'W, 250 m, *Løjtnant & Molau 13489* (AAU, GB). Illustrations based on: *Kegel 741* (upper right), *Krause s.n.* (NY) - lower right, *Allioni 625* (H) - upper left, *Spruce 748* (NY) - lower left.

In Ecuador known only from the above cited provinces; at elevations from 245 m to ca. 1000 m.

Lepidopilum grevilleanum (Tayl.) Mitt. has been considered a synonym of *L. affine* but that taxon is a distinct, rare species from the coastal premontane forests of Ecuador, known only from the provinces of Esmeraldas and Los Ríos (Churchill, 1992). The collections by Harling from Morona-Santiago were originally determined as *Lepidopilum armatum* Mitt. by Crum (1957).

Lepidopilum arcuatum Mitt., J. Linn. Soc., Bot. 12: 374. 1869. (Fig. 11A-C).

Pastaza: Río Bobonaza, 460 m, *Spruce 733* (type, NY; Mitten, 1869). Illustration based on *Spruce 733* (NY).

Lepidopilum brevipes Mitt., J. Linn. Soc., Bot. 12: 376. 1869. (Fig. 10E-H).

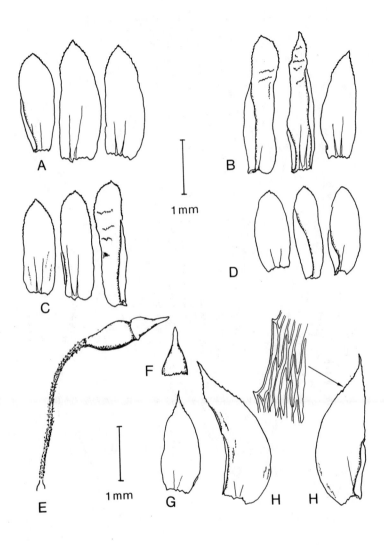

Figure 10. A-D. *Lepidopilum affine,* A-D. illustrating variation in leaf shapes; E-H. *Lepidopilum brevipes,* E. sporophyte, F. calyptra, G. median leaf, H. lateral leaves.

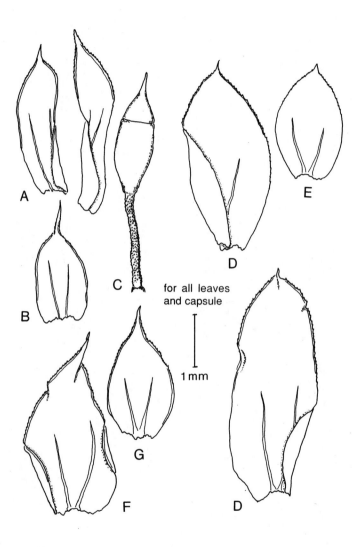

Figure 11. A-C. *Lepidopilum arcuatum*, A. lateral leaves, B. median leaf, C. sporophyte; D-E. *Lepidopilum curvifolium*, lateral leaves, E. median leaf; F-G. *Lepidopilum polytrichoides*, F. lateral leaf, G. median leaf.

Napo: Ca. 6.7 km S of Tena on road to Puyo, 600 m, *Churchill & Sastre-De Jesús 13654* (NY). Illustration based on *Churchill et al. 13976* (NY) from Colombia.

In Ecuador known from Azuay, Chimborazo and Pastaza; at elevations from 600–1300 m, possibly to 2000 m or more.

Lepidopilum curvifolium Mitt., J. Linn. Soc., Bot. 12: 374. 1869. (Fig. 11D-E).

Napo: Añangu, ca. 75 km east of Coca, ca. 00°32'S, 76°23'W, 245–325 m, *Churchill & Sastre-De Jesús 13800, 13819* (NY; Churchill et al., 1992). **Pastaza:** Río Bobonaza, 365–610 m, *Spruce 724* (type NY; Mitten, 1869). Illustration based on *Spruce 724* (NY) - upper, *Buchtier s.n.* (H) - lower.

In Ecuador also known from Morona-Santiago; at elevations to 800 m or more.

Lepidopilum pallido-nitens (C. Müll.) Par., Ind. Bryol. Suppl. 223. 1900. (Fig. 12A-B). Synonym: *Lepidopilum attenuatum* Bartr.

Napo: Shinguipino, between Río Napo and Tena, ca. 460 m, *Grubb 2939a* (type of *L. attenuatum*, BM), *2946a* (Bartram, 1964). Illustration based on *Grubb 2939a* (US).

Lepidopilum polytrichoides (Hedw.) Brid., Bryol. Univ. 2: 269. 1827. (Fig. 11F-G).

Napo: Shinguipino, between Río Napo and Tena, ca. 460 m, *Grubb et al. 2923, 2923b, 2923b, 2950* (Bartram, 1964); Hotel Jaguar on Río Napo, ca. 50 km below Puerto Misahualli, ca. 400 m, *Steere E-95* (NY); 2–5 km W of San Pablo de los Secoyas, 00°15'S, 77°21'W, 300 m, *Brandbyge et al. 32480* (AAU); Añangu, ca. 75 km east of Coca, ca. 00°32'S, 76°23'W, 245–325 m, *Churchill & Sastre-De Jesús 13744* (NY; Churchill et al., 1992); Río Suno, 00°42'S, 77°10'W, 400 m, *Holm-Nielsen & Jeppesen 887* (AAU, GB), *888* (AAU); Yuralpa, 00°55'S, 77°21'W, 440 m, *Holm-Nielsen & Jeppesen 944*, (AAU, GB; Robinson et al., 1971). **Pastaza:** Río Bobonaza, 365 m, *Spruce 713* (NY; Mitten, 1869). Illustration based on *Swartz s.n.* (BM) from Jamaica.

In Ecuador known also from Azuay, Los Ríos and Morona-Santiago; at elvations from 245–1000 m or more.

Lepidopilum radicale Mitt., J. Linn. Soc., Bot. 12: 378. 1869.

Pastaza: Río Bobonaza, 395 m, *Spruce 769, 770* (NY; Mitten, 1869).

Lepidopilum scabrisetum (Schwaegr.) Steere, Bryologist 51: 140. 1948.

Napo: Shinguipino, between Río Napo and Tena, ca. 460 m, *Grubb et al. 2955* (Bartram, 1964); Hotel Jaguar on Río Napo, ca. 50 km below Puerto Misahualli, ca. 400 m, *Steere E-114* (NY); Añangu, ca. 75 km east of Coca, ca. 00°32'S, 76°23'W, 245–325 m, *Churchill & Sastre-De Jesús 13791-d* (NY; Churchill et al., 1992).

In Ecuador known from Esmeraldas, Pastaza and Pichincha; at elevations from 245–1800 m.

Lepidopilum surinamense C. Müll., Linnnaea 21: 193. 1848. (Fig. 12C-D). Synonym: *Lepidopilum flexifolium* (C. Müll.) Mitt.

Monora-Santiago: Yurupaz, 600 m, *Harling 2241b* (S; collection with *L. affine*). **Napo:** Km 15 S of Coca on Coca-Armenia Vieja road, ca. 250 m, *Fransén 47* (GB); Añangu, ca. 75 km

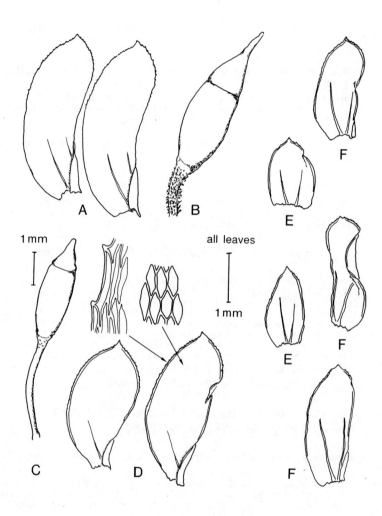

Figure 12. A-B. *Lepidopilum pallido-nitens*, A. lateral leaves, B. capsule; C-D. *Lepidopilum surinamense*, C. sporophyte, D. leaves; E-F. *Lepidopilum tortifolium*, E. lateral and F. median (left side) leaf variation.

east of Coca, ca. 00°32'S, 76°23'W, 245–325 m, *Churchill & Sastre-De Jesús 13787, 13791-c, 13809* (AAU, NY; Churchill *et al.*, 1992). **Pastaza:** Río Bobonaza, 365 m, *Spruce 721* (NY; Mitten, 1869). **Sucumbíos:** Reserva Faunística Cuyabeno, near Laguna Grande, 00°00', 76°12'W, 265 m, *Balslev 84916* (AAU). Illustrations based on: *Kegal 1406* (H) - leaves, *Spruce 719* (NY) - capsule.

In Ecuador known from Chimborazo, Esmeraldas and Pichincha; at elevations to ca. 800 m.

Lepidopilum tortifolium Mitt., J. Linn. Soc., Bot. 12: 374. 1869. (Fig. 12E-F). Synonym: *Lepidopilum crispifolium* Bartr.

Napo: Shinguipino, between Río Napo and Tena, ca. 460 m, *Grubb et al. 2939, 2942a, 2947* (BM, type of *L. crispifolium*; Bartram, 1964). Río Aguarico, E of the mouth of Río Cuyabeno, 00°16'S, 75°54'W, 200 m, *Holm-Nielsen et al. 21494* (AAU). **Pastaza:** Río Bobonaza, 365 m, *Spruce 783* (NY; Mitten, 1869). Illustrations based on: upper - *Grubb et al. 2947* (BM), upper and middle - *Weir 147* (NY) from Colombia.

In Ecuador known also from Esmeraldas; at elevations from 105–460 m or more.

Pilotrichidium Besch.

Pilotrichidium callicostatum (C. Müll.) Jaeg., Ber. St. Gall. Naturw. Ges. 1875–76: 357. 1877. (Fig. 8A).

Plants pale to dark green or blackish-green, primary stems creeping; secondary stems spreading, leaves contorted when dry, ovate-oblong or oblong, 1.3–1.8 mm long, 0.5–0.7 mm wide, apex truncate or obtuse-rouned, margins appearing finely serrulate by projecting papillae, costae strong, double, parallel, ending just below leaf apex, terminating in a projecting blunt tooth, median cells irregularly quadrate to hexagonal, thick-walled, pluripapillose, 2–3 papillae over cell lumen; dioicous; setae elongate, capsules erect, urn oblong, exostome striate, weakly furrowed, opercula rostrate.

In Ecuador known only from the locality given below. It is likely to be more common in the upper reaches of the Amazonas and premontane region of Ecuador; generally growing over logs or on rocks in streams. Reference: Allen and Crosby (1986a).

Napo: Shinguipino, between Río Napo and Tena, ca. 460 m, *Grubb et al. 2945* (Bartram, 1964). Illustration based on *Callejas et al. 2799* (NY) from Colombia.

CALYMPERACEAE

Plants dark to glossy green, often crispate, forming tufts or solitary; leaves usually sheathing at base, base often expanded, ovate or oblong, distally lingulate to narrowly lanceolate, acuminate to acute, or rounded to subtruncate, margins entire or more commonly serrate or ciliate, teeth single, double or more, costae single, mostly percurrent, distal lamina (limb) cells smooth to more often papillose, sheathing basal cells strongly differentiated, region adjacent to costa of enlarged, clear cells (cancellinae), outer cells beyond the cancellinae often with a differentiated band of linear cells (teniolae) usually separated by isodiametric cells or margins brodered by linear cells; propagula are often present on the upper lamina or distally with the apex highly modified; setae elongate, smooth; capsules erect, urn cylindrical, peristome absent or present with 16 papillose teeth, opercula rostrate; calyptrae cucullate or campanulate and often persistent (clasping at base) with few to several longitudial slits.

Calymperaceae are typically epiphytic in shaded places, on trunks of treelets, trees and palms, occasionally epiphyllous. The family is a common and characteristic element in the lowlands of the Neotropics. Both the biology and morphology of this family are exceptionally interesting, and the plants are very attractive when viewed under the microscope. Further species should be discovered in Amazonas than is presently reflected in this treatment. References: Reese (1983); distribution and evolution of our two genera have been treated by Reese (1987a, 1987b).

Key to the Genera

1. Leaves with *either* a intramarginal band of elongate cells, *or* distal lamina with transversely elongate or subrectangular cells; calyptra persistent, campanulate with longitudinal slits; peristome usually present   *Calymperes*
1. Leaves lacking a intramarginal band of elongate cells, margins however limbate or elimbate, distal cells not transversely elongate, mostly quadrate or quadrate-rounded; calyptra deciduous, cucullate; peristome always absent   *Syrrhopodon*

Clave para los Géneros

1. Hojas ya sea con una banda intramarginal de células elongadas, o la lámina con células distales transversalmente elongadas o subrectangulares; caliptra persistente, campanulada con hendiduras longitudinales; perístoma usualmente presente   *Calymperes*
1. Hojas sin banda intramarginal de células elongadas, márgenes sin embargo, bordeados o sin borde; células distales no transversalmente elongadas, la mayoría cuadradas o cuadrado-redondeadas; caliptra decidua, cuculada; perístoma siempre ausente   *Syrrhopodon*

Calymperes Sw. in Web.

Leaves obovate to oblong at base, sheathing, distally lanceolate to linear, apex narrow and blunt, margin finely serrate along sheathing base, costae subpercurrent, distal lamina cells irregularly quadrate-rounded or transversely elongate, papillose or mamillose, cancellinae large, adjoining region with cells similar to those of distal lamina, with or without an intramarginal band of linear cells; peristome absent; calyptrae campanulate, plicate, usually twisted at base, at maturity persisting with longitudinal slits.

In Ecuador four species are recorded, however further species are to be expected. Reference: Reese (1961).

Key to the Species
1. Distal lamina cells transversely elongate; leaves elongate, 10–15 mm or more long   *C. lonchophyllum*
1. Distal lamina cells not transversely elongate; leaves short, to 5 mm long   **2.**
   2. Shoulder margin entire or minutely dentate   *C. afzelii*
   2. Shoulder margin serrulate to serrate   *C. erosum*

Clave para las Especies
1. Células distales de la lámina tranversalmente elongadas; hojas elongadas, 10–15 mm o más de longitud   *C. lonchophyllum*
1. Células distales de la lámina no transversalmente elongadas; hojas cortas, hasta 5 mm de longitud   **2.**
   2. Márgen de la región de los hombros entero o finamente dentado
                                                                 C. afzelii
   2. Márgen de la región de los hombros serrulado hasta serrado
                                                                 C. erosum

Calymperes afzelii Sw., Jahrb. Gewächsk. 1: 3. 1818. Synonym. *Calymperes donnellii* Aust.
Sucumbiós: Reserva Faunística Cuyabeno, N of Laguna Grande, 00°01'N, 76°11'W, 265 m, *Heikkinen RH-1990–181* (NY).
Reported from Esmeraldas and the Galapagos Islands.

Calymperes erosum C. Müll., Linnaea 21: 182. 1848. (Fig. 14A).
Napo: Ca. 15 km S of Coca on Coca-Armenia Vieja road, ca. 250 m, *Fransén 47* (GB, with collection of *Lepidopilum surinamense*). Illustration based on *Fransén 47*.
Previously reported from the Galapagos Islands.

Calymperes lonchophyllum Schwaegr., Spec. Musc. Suppl. 1(2): 333. 98. 1816. (Fig. 13A-C).

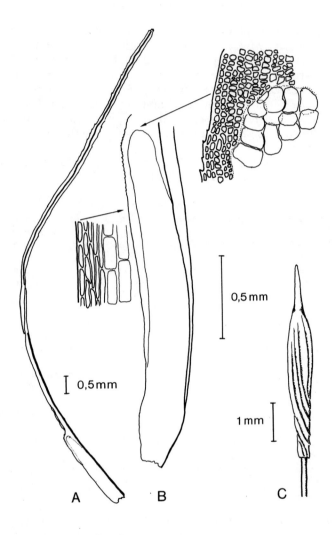

Figure 13. A-C. *Calymperes lonchophyllum,* A. leaf, B. leaf base, C. capsule with attached calyptra.

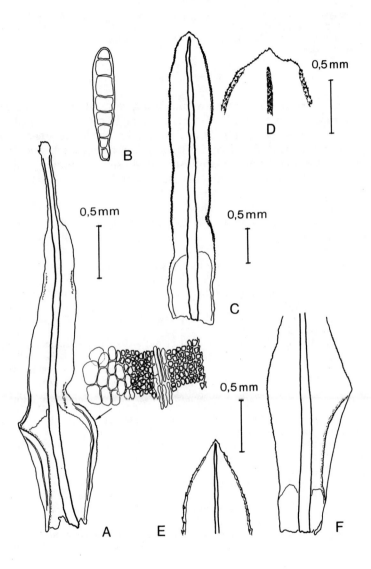

Figure 14. A. *Calymperes erosum*, A. leaf illustrating intramarginal band; B-D. *Syrrhopodon cryptocarpos*, B. propagula, C. leaf, D. leaf apex; E-F. *Syrrhopodon rigidus*, E. leaf apex, F. leaf base.

Napo: 15 km south of Coca, 250 m, *Fransén 49* (GB); Añangu, ca. 75 km east of Coca, ca.
00°32'S, 76°23'W, 245–325 m, *Churchill & Sastre-De Jesús 13877* (AAU, NY; Churchill *et al.*,
1992); *Brako 5165* (NY). **Pastaza:** Curaray, 200 m, *Andersson 882* (AAU, GB). Illustration
based on *Andersson 882*.

Syrrhopodon Schwaegr.

Leaves with oblong, obovate sheathing base, distally lingulate,
lanceolate or linear, apex often blunt, acute to rounded, margins of
sheathing base serrulate, serrate or ciliate, border of linear cells
present or absent, distal lamina cells quadrate-rounded, smooth to
more often papillose or papillose-mamillose, cancellinae large or
reduced and inconspicuous; peristome single, teeth 16, papillose;
calyptrae cucullate, deciduous.

Syrrhopodon contains about 11 species in Ecuador. Further
species of this genus should be expected in Amazonas. References:
Reese (1977, 1978).

Key to the Species

1. Leaf margins lacking a border (elimbate)   **2.**
   2. Marginal teeth of leaf in 3 or more irregular rows; leaf apex
     ±truncate-apiculate   *S. cryptocarpos*
   2. Marginal teeth of leaf, if present, in 2 rows; leaf apex acute to
     rounded   **3.**
      3. Cancellinae ending far below leaf shoulder, extending about
        1/3 distance of leaf shoulder length   *S. hornschuchii*
      3. Cancellinae extending into leaf shoulder at broadest point   **4.**
        4. Leaf shoulder broadly obovate, margins weakly
          serrulate at shoulder   *S. incompletus*
        4. Leaf shoulder ovate, shoulder margins rather strongly
          serrate with at least some teeth recurved   *S. rigidus*
1. Leaf margins completely or incompletely bordered (limbate)   **5.**
   5. Plants small, leaves to 2 mm long, apex rounded   *S. ligulatus*
   5. Plants medium sized, leaves 3 mm or longer, apex acute or
     acuminate   **6.**
      6. Margins of ovate leaf base ciliate   *S. leprieurii*
      6. Margins of leaf base entire, not ciliate   **7.**
        7. Distal lamina broad, 1.0–1.5 mm wide; apices acute to
          broadly acuminate   *S. parasiticus*
        7. Distal lamina narrow, to 0.5 mm wide; apices narrowly
          acuminate   *S. prolifer* var. *acanthoneuros**

Clave para las Especies

1. Márgen de la hoja sin borde (elimbada)   **2.**
   2. Dientes marginales de la hoja en 3 o más filas irregulares; ápice de
     la hoja ±truncado-apiculado   *S. cryptocarpos*

2. Dientes marginales de la hoja, si están presentes, en 2 filas; ápice de la hoja agudo hasta redondeado **3.**

3. Cancelina terminando mucho mas abajo de la región de los hombros en la hoja, extendiendose cerca 1/3 parte de la longitud de la región de los hombros *S. hornschuchii*

3. Cancelina extendiendose hasta la región de los hombros en la hoja **4.**

4. Región de los hombros ampliamente obovada, margenes levemente serrulados *S. incompletus*

4. Región de los hombros ovada, márgenes fuertemente serradas con almenos algunos dientes recurvados *S. rigidus*

1. Márgenes de la hoja parcial o completamente bordeados (limbada) **5.**

5. Plantas pequeñas, hojas hasta de 2 mm de longitud, ápice redondeado *S. ligulatus*

5. Plantas de tamaño mediano, hojas de 3 mm de longitud o más, ápice agudo o acuminado **6.**

6. Margenes de la base de la hoja ovada, ciliados *S. leprieurii*

6. Márgenes de la base de la hoja, entera, no ciliados **7.**

7. Lámina distal ancha, 1.0–1.5 mm de ancho; ápice agudo hasta ampliamente acuminado *S. parasiticus*

7. Lámina distal estrecha, hasta 0.5 mm de ancho; ápice estrechamente acuminado *S. prolifer var. acanthoneuros**

Syrrhopodon cryptocarpos Dozy & Molk., Natuurk. Verh. Holl. Maatsch. Wet. 10(3): 14. 7. 1854. (Fig. 14B-D).

Napo: Añangu, ca. 75 km east of Coca, ca. 00°32'S, 76°23'W, 245–325 m, *Churchill & Sastre-De Jesús 13790* (NY; Churchill *et al.*, 1992). **Pastaza:** Montalvo, trail to Chiriboga, on Río Bobonaza, 02°05'S, 76°58'W, 300–350 m, *Øllgaard et al. 35512* (AAU); Oil exploration camp Chichirota, along Río Bobonaza, ca. 02°22'S, 76°40'W, ca. 300 m, *Øllgaard et al. 35289* (AAU). **Sucumbíos:** Reserva Faunística Cuyabeno, near Laguna Grande, 00°00', 76°12'W, 265 m, *Balslev 84915* (AAU); *Heikkinen RH-1990-135* (NY). Illustration based on *Churchill & Sastre-De Jesús 13790*.

In Ecuador reported also from Morona-Santiago; at elevations from 245–800 m.

Syrrhopodon hornschuchii Mart., Fl. Brasiliensis 1(2): 6. 1840.

Sucumbíos: Reserva Faunística Cuyabeno, N of Laguna Grande, 00°01'N, 76°11'W, 265 m, *Heikkinen RH-1990-64* (NY).

Also known from Morona-Santiago (Reese, 1993).

Syrrhopodon incompletus Schwaegr., Spec. Musc. Suppl. 2(1): 119. 1824. var. *incompletus* (Fig. 15D).

Napo: Añangu, ca. 75 km east of Coca, ca. 00°32'S, 76°23'W, 245–325 m, *Churchill & Sastre-De Jesús 13801* (NY; Churchill *et al.*, 1992). **Sucumbíos:** Reserva Faunística Cuyabeno, N of Laguna Grande, 00°01'N, 76°11'W, 265 m, *Heikkinen RH-1990-165b* (NY). Illustration based on *Churchill & Sastre-De Jesús 13801*.

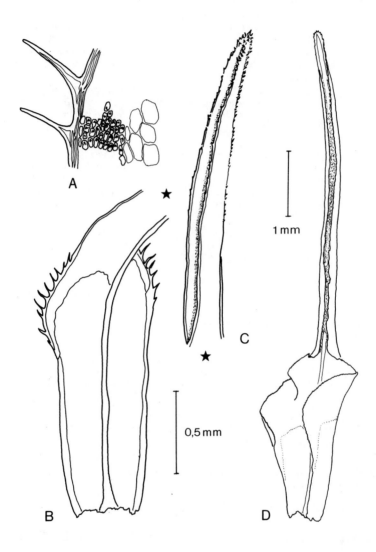

Figure 15. A-C. *Syrrhopodon leprieurii*, A. cell pattern along distal leaf shoulder, B. leaf base, C. leaf apex; D. *Syrrhopodon incompletus*, D. leaf.

Syrrhopodon leprieurii Mont., Ann. Sc. Nat. Bot. Sér. 2, 2: 379. 1834. (Fig. 15A-C).

Napo: Río Aguarico, Tangoy, 1 hour downstream from Zancudo, 00°34'S, 75°27'W, 300 m, *Holm-Nielsen et al. 20099* (AAU). **Pastaza:** Lorocachi, along Río Curaray, 01°38'S, 75°58'W, 200 m, *Jaramillo et al. 31175* (AAU). Illustration based on *Jaramillo et al. 31175*.

At elevations from 200–1540 m.

Syrrhopodon ligulatus Mont., Syll. Gen. Sp. Crypt. 47. 1856. (Fig. 16A-B).

Sucumbíos: Reserva Faunística Cuyabeno, N of Laguna Grande, 00°01'N, 76°11'W, 265 m, *Heikkinen RH-1990-137p.p.* (NY). Illustration based on *Heikkinen RH-1990-137p.p.*

New to Ecuador. This species, readily overlooked due to its small size, is widespread but rarely collected in the Neotropics.

Syrrhopodon parasiticus (Brid.) Besch., Ann. Sci. Nat. Bot. 8, 1: 298. 1895. (Fig. 16C-D).

Sucumbíos: Reserva Faunística Cuyabeno, N of Laguna Grande, 00°01'N, 76°11'W, 265 m, *Heikkinen RH-1990-21a, RH-1990-28* (NY). Illustration based on *Heikkinen RH-1990-21a*.

Reported from the Galapagos Islands. In addition to occurring on treelet branches, this species is also epiphyllous, often on the margin of leaves.

Syrrhopodon rigidus Hook. & Grev., Edinburgh J. Sci. 3: 226. 1825. (Fig. 14E-F).

Napo: Near Tena, 6 km along Río Pano, 00° 58'S, 77°52'W, 600 m, *Holm-Nielsen & Jeppesen 698* (AAU; Robinson *et al.*, 1971). **Pastaza:** Curaray, Valle de la Muerte, 240 m, 01°25'S, 76°52'W, *Holm-Nielsen et al. 22420* (AAU). Illustration based on *Holm-Nielson et al. 22420*.

In Ecuador reported also from Morona-Santiago; at elevations from 240–1000 m.

DALTONIACEAE

Leskeodon Broth.

Leskeodon andicola (Mitt.) Broth., Nat. Pfl. 1(3): 926. 1907 (Fig. 17A)

Plants pale whitish to olive green, delicate and lax, forming mats, stems spreading or subascending, leaves complanate, differentiated between lateral and median, lateral leaves oblong, 1.5–2 mm long, 0.6–0.8 mm wide, short apiculate, often twisted, margins entire, costae 2/3–3/4 lamina length, median cells hexagonal-rounded, basal cells irregularly rectangular, lax, marginal cells forming a border, 1–2 rows of linear cells; autoicous; setae to 4 mm long, distally roughened; capsules erect to suberect, urn ovoid short cylindrical, peristome double, papillose, not

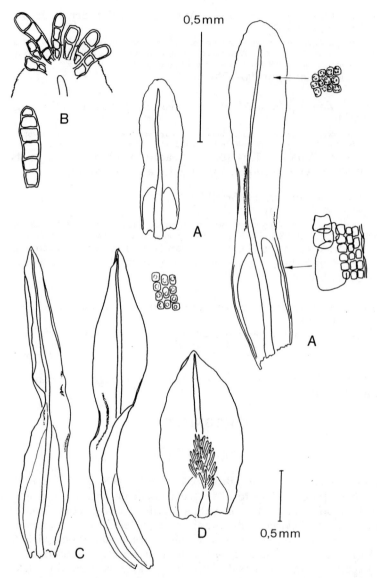

Figure 16. A-B. *Syrrhopodon ligulatus*, A. leaves, B. leaf tip with propagula; C-D. *Syrrhopodon parasiticus*, C. leaves, D. comal leaf with propagula.

furrowed; opercula short rostrate; calyptrae mitrate, hairy, at base fringed with hairs.

Rarely abundant and easily overlooked, *Leskeodon* is found in shaded places, often on trunks or branches of treelets and shrubs, and on logs. Two species are presently known for Ecuador. Traditionally *Leskeodon* has been placed in the Hookeriaceae; in the present treatment it is placed in the Daltoniaceae along with *Calyptrochaeta* and *Daltonia*, both of which occur in montane forests of Ecuador. Reference: Buck (1987a).

Napo: Shinguipino, between Río Napo and Tena, ca. 460 m, *Grubb et al. 2942* (Bartram, 1964). **Pastaza:** Río Bobonaza, 305–460 m, *Spruce 572* (NY; Mitten, 1869); Montalvo, south side of Río Bobonaza, 0–2 km SW of military camp, 02°05'S, 76°58'W, 250–300 m, *Løjtnant & Molau 13551* (AAU, GB). **Sucumbíos:** Reserva Faunística Cuyabeno, near Laguna Grande, 00°00', 76°12'W, 265 m, *Balslev 84919* (AAU). Illustration based on *Balslev 84919*.

Reported at elevations from 250–925 m.

DICRANACEAE

Plants erect, often forming loose to dense tufts, occasionally solitary; leaves loosely erect to falcate-secund, ovate-lanceolate to narrowly lanceolate, costae single, median cells mostly quadrate to rectangular, smooth, bulging or papillose, alar region undifferentiated or differentiated, basal cells often elongate; sporophytes terminal, setae elongate, smooth, capsules erect or curved, urn ovoid to ovoid-cylindrical, peristome single, teeth 16 and often partially to fully divided, papillose or variously striate below; opercula rostrate; calyptrae cucullate, smooth, rarely hairy at base.

In the tropical lowlands species of this family are found as epiphytes in shaded places and on soil, often in exposed and disturbed places. *Campylopus*, one of the largest genera in Ecuador (30 species) and the Neotropics, should possibly be found in our area. Reference: Frahm (1991).

Key to the Genera

1. Costae broad, 1/2 or more width at leaf base   *Campylopus**
1. Costae narrow, 1/3 or less width at leaf base   **2.**
    2. Leaf bordered, marginal cells linear   *Leucoloma*
    2. Leaf not bordered, marginal cells ±similar to lamina cells   **3.**
        3. Alar cells differentiated, walls porose; perichaetial leaves strongly clasping and ±enveloping 1/2 or more the length of the seta   *Holomitrium*
        3. Alar cells not differentiated, walls not porose; perichaetial leaves short   **4.**

4. Capsules with a distinct elongate neck, ca. 2 time urn length; leaves not secund   *Trematodon*
4. Capsules lacking an elongate neck below urn, if present short; leaves ±secund   **5.**
    5. Leaf margins entire, not serrate or dentate except at apex; setae to 8.5 mm long; peristome elongate, pitted, divided   *Dicranella*
    5. Leaf margins irregularly serrate or dentate; setae to 5 mm long; peristome short, papillose, undivided
                                           Microdus

Clave para los Géneros

1. Costas anchas, 1/2 o más del ancho de la base de la hoja *Campylopus**
1. Costas angostas, 1/3 parte o menos del ancho de la base de la hoja   **2.**
  2. Hoja bordeada, células marginales lineares   *Leucolma*
  2. Hoja no bordeada, células marginales ±similares a las células de la lámina   **3.**
    3. Células alares diferenciadas, paredes porosas; hojas periqueciales fuertemente convolutas y ±envolviendo la 1/2 o más de la longitud de la seta   *Holomitrium*
    3. Células alares no diferenciadas, paredes no porosas; hojas periqueciales cortas   **4.**
      4. Cápsulas con cuello claramente elongado, ca. 2 veces la longitud de la urna, hojas secundiformes *Trematodon*
      4. Cápsulas sin cuello elongado inferior a la urna; hojas ±secundiformes   **5.**
        5. Márgenes de la hoja enteros, no serrados o dentados excepto hacia el ápice; setas hasta 8.5 mm de longitud; perístoma elongado, dividido
                                      Dicranella
        5. Márgenes de la hoja irregularmente serrados o dentados; setas hasta 5 mmm de longitud; perístoma corto, papiloso, no dividido *Microdus*

Dicranella (C. Müll.) Schimp.

Dicranella hilariana (Mont.) Mitt., J. Linn. Soc., Bot. 12: 31. 1869. (Fig. 18C-F).

Plants somewhat small, to 1.5 cm tall; leaves often secund, narrowly lanceolate from an expanded base, 1.2–2.5 mm long, ca. 0.2–0.3 mm wide at the base, ±channeled above, margins entire, apex distinctly serrate-toothed, costae subpercurrent, median cells quadrate to more commonly rectangular, smooth, basal cells rectangular, somewhat larger but not differentiated; setae to 10 mm long or more, capsule ovoid, peristome divided to half or more, opercula long rostrate.

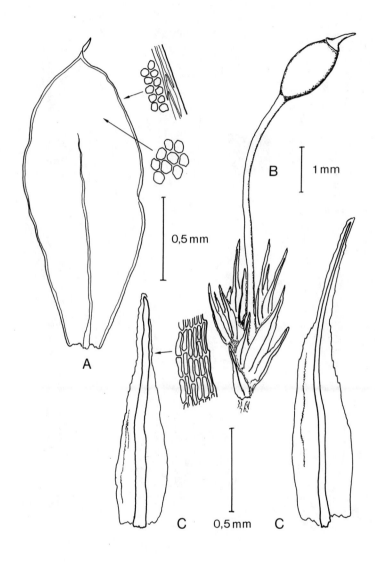

Figure 17. A. *Leskeodon andicola*, A. leaf; B-C. *Microdus exiguus*, B. habit, C. leaves.

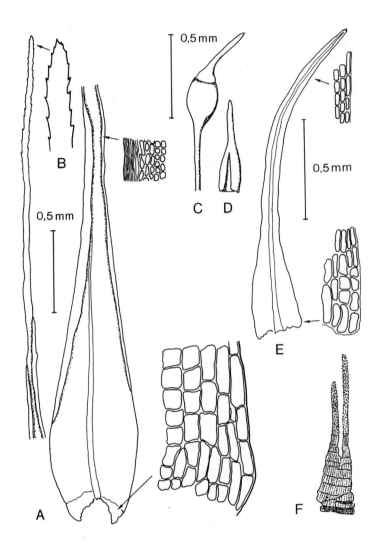

Figure 18. A-B. *Leucoloma serrulatum*, A. leaf, B. leaf apex; C-F. *Dicranella hilariana*, C. capsule, D. calyptra, E. leaf, F. single divided peristome tooth.

Dicranella is commonly found on soil in exposed places, often in disturbed sites. Approximately nine species are recorded for Ecuador.

Napo: Coca-Curaray road, km 40, ca. 250 m, *Fransén 3* (AAU, GB). Illustration based on *Fransén 3.*

In Ecuador reported also from Morona-Santiago; from elevations at 250 m to possibly 1000 m or more.

Holomitrium Brid.

Holomitrium arboreum Mitt., J. Linn. Soc., Bot. 12: 58. 1869.

Plants dark green to ±blackish-green, forming tall turfs; leaves crispate, obovate-lanceolate, base obovate, rather abruptly narrowed above, ca. 4–5 mm long, 0.5–0.8 mm wide, apex acuminate, margins above obovate base sharply serrate, costae strong, percurrent, median cells quadrate- to short rectangular-rounded, basal and insertion cells rectangular and porose, golden-brown; perichaetial leaves enveloping 1/2 or more of seta length, setae erect, 8–12 mm long, capsule erect, urn ovoid-cylindrical, peristome single or divided, papillose; opercula rostrate; calyptrae cucullate.

This species should be expected on trunks or branches of shrubs or treelets in somewhat exposed places but more likely in the upper canopy in Amazonian primary forests. *Holomitrium* contains seven species in Ecuador, found mostly at higher elevations in montane forests. Reference: Hegewald (1978).

Pastaza: Río Bobonaza & Canelos, 610 m, *Spruce 22b* (NY; Mitten, 1869).

Leucoloma Brid.

Leucoloma serrulatum Brid., Bryol. Univ. 2: 752. 1827. (Fig. 18A-B).

Plants forming turfs, several branched, somewhat secund, leaves ovate-long lanceolate or -subulate, 5–6 mm long, ca. 0.5 mm wide, concave, acute, channeled, margins 1/4–1/3 serrulate distally, costae exurrent, median cells quadrate to short rectangular, weakly papillose, marginal cells forming a border of linear cells, basal cells enlarged, oblong-rectangular, ±golden-yellow; dioicous; setae elongate, capsules erect, urn obovoid to cylindrical, peristome teeth 16, divided to near base, papillose distally; opercula rostrate.

Leucoloma is commonly found in shaded places as an epiphyte on trees or treelets; three species of this genus are recorded for Ecuador.

Napo: near Tena, 6 km along Río Pano, 00° 58'S, 77°52'W, 600 m, *Holm-Nielsen & Jeppesen* 692 (AAU; Robinson *et al.*, 1971). Illustration based on *Holm-Nielsen & Jeppesen* 692.

Microdus Schimp. in Besch.

Microdus exiguus (Schwaegr.) Besch. *in* Par., Ind. Bryol. 803. 1897. (Fig. 17B-C).
Plants small; leaves ±falcate-secund, ovate-subulate, 1.2–2.2 mm long, margins plane to slightly recurved below, irregularly dentate distally, costae sub- or percurrent, median and basal cells short to long rectangular, smooth, alar region not differenitated; perichaetial leaves longer than stem leaves; setae 4–5 mm long, erect to slightly curved; urn oblong-oval, ca. 1 mm long, peristome short, undivided, papillose; opercula rostrate, ± oblique.
This inconspicuous species is generally found on exposed soil. Three species of *Microdus* are presently known from Ecuador.
Pastaza: Río Bobonaza, 365 m, *Spruce 39* (NY; Mitten, 1869). Illustration based on *Reineck & Czermak 27* (GB) from Brazil.

Trematodon Michx.

Trematodon humilis Mitt., J. Linn. Soc., Bot. 12: 47. 1869.
Plants small; stems simple; leaves ±sheathing stem, narrowly lanceolate from an oblong-ovate base, ca. 1.4 mm long, apex obtuse, serrulate or dentate, median cells quadrate to short rectangular, smooth; perichaetial leaves long subulate, to 2.5 mm long; setae elongate, to 12 mm long, smooth and twisted; capsule suberect, urn oblong-cylindrical, neck distinct, 2 or more times longer than capsule, peristome single, teeth 16; opercula long rostrate, oblique; calyptrae cucullate.
Only a single species of *Trematodon* is presently known for Ecuador. The genus is possibly best placed in the family Bruchiaceae (as in the Ecuadorean checklist, see Appendix 1).
Pastaza: Río Bobonaza, 365 m, *Spruce 44b* (NY; Mitten, 1869).

FISSIDENTACEAE

Fissidens Hedw.

Plants very small to medium sized, erect and forming short turfs or occurring solitary, rarely procumbent (*F. hydropogon*); leaves 2-ranked, median and upper leaves with lower basal vaginant laminae (sheath), distally with an extended dorsal and ventral lamina; costae percurrent to short excurrent; margins or intramargins limbate or elimbate; cells smooth or uni- or pluripapillose; setae terminal, elongate, rarely short (*F. hydropogon*), erect or variously curved; peristome single, teeth 16, striate or papillose; opercula short to long rostrate; calyptrae cucullate or short mitrate.

Fissidens is commonly found in shaded places, on soil and rocks, occasionally on trunks of treelets and trees, rarely aquatic (*F. hydropogon*). A large genus with numerous species in the Neotropics; Ecuador possibly contains more than 27 species. Several more species other than those given here should be expected in the Amazonas. Reference: Pursell (1984; pers. comm.).

Key to the Species

1. Plants lax, trailing, irregularly pinnately branched; seta short; aquatic
   F. hydropogon
1. Plants erect, rarely branched; seta elongate; mostly terrestrial or epiphytic **2.**
   2. Leaves limbate, bordered with long linear cells throughout or partially **3.**
      3. Leaf border confined to vaginant laminae on stem leaves or on perichaetial leaves; cells distinctly papillose **4.**
         4. Cells unipapillose, papillae sharply pointed; border of linear cells confined to perichaetial leaves; vaginant lamina ca. 1/2 lamina length *F. diplodus*
         4. Cells uni- or pluripapillose, papillae not sharply pointed; border confined to vaginant lamina or on perichaetia of stem leaves; vaginant lamina 2/3 or more length of lamina **5.**
            5. Leaf cells unipapillose *F. intermedius*
            5. Leaf cells pluripapillose *F. intramarginatus*
      3. Leaf border extending and merging with apex; cells smooth **6.**
         6. Costa ending several cells below apex; median cells enlarged, thin-walled, lax *F. mollis*
         6. Costa ending in apex; median cells smaller, thick-walled, not lax except in lower juxtacostal part of vaginant laminae *F. zollingeri*
   2. Leaves elimbate, marginal cells isodiametric and ±similar to inner lamina cells **7.**
      7. Leaf cells rather large and lax *F. inaequalis*
      7. Leaf cells small and firm-walled, not appearing lax **8.**
         8. Leaf cells unipapillose **9.**
            9. Cell papillae not sharply pointed; costa ending several cells below apex *F. steerei*

9.   Cell papillae sharply pointed; costa ±percurrent
                                              F. diplodus
8.   Leaf cells mamillose or bulging   **10.**
   10.   Plants ca. 1–2 cm tall; ca. 18 pairs of leaves or more,
          leaves 2–2.5 mm long   *F. asplenioides*
   10.   Plants ca. 0.5 cm tall; 3–10 pairs of leaves, rarely
          more, leaves 1–2 mm long   *F. prionodes*

Clave para las Especies

1. Plantas laxas, rastreras, irregular y pinnadamente ramificadas; seta corta; acuáticas   *F. hydropogon*
1. Plantas erectas, raramente ramificadas; seta elongada; la mayoría terrestres o epífitas   **2.**
   2.   Hojas bordeadas completa o parcialmente por células lineares largas   **3.**
      3.   Borde de la hoja confinado a la lámina vaginante de las hojas del tallo o en las hojas periqueciales; células distintivamente papilosas   **4.**
         4.   Células unipapilosas, papilas fuertemente puntiagudas; borde de células lineares confinado a las hojas periqueciales; lámina vaginante ca. 1/2 de la longitud de la lámina   *F. diplodus*
         4.   Células uni-o pluripapilosas, papilas levemente puntiagudas; borde confinado a la lámina vaginante o a las hojas periqueciales; lámina vaginante 2/3 partes o más de la longitud de la lámina   **5.**
            5.   Células de la hoja unipapilosas   *F. intermedius*
            5.   Células de la hoja pluripapilosas
                                              F. intramarginatus
      3.   Borde de la hoja extendiendose y fusionandose con el ápice; células lisas   **6.**
         6.   Costa terminando varias células por debajo del ápice; célulasmedias agrandadas, paredes delgadas, laxas
                                                    F. mollis
         6.   Costa terminando en el ápice; células medias de menor tamaño, paredes engrosadas, no laxas, excepto en la parte inferior juxtacostal de la lámina vaginante
                                                  F. zollingeri
   2.   Hojas no bordeadas, células marginales isodiamétricas y ±similares a las células al interior de la lámina   **7.**
      7.   Células de la hoja un poco grandes y laxas   *F. inaequalis*
      7.   Células de la hoja pequeñas y con paredes firmes, no luciendo laxas   **8.**
         8.   Células de la hoja unipapilosas   **9.**
            9.   Papilas celulares no fuertemente puntiagudas; costa terminando varias células más abajo del ápice   *F. steerei*

9. Papilas celulares fuertemente puntiagudas; costa
   ±percurrente   *F. diplodus*
8. Células de la hoja mamilosas o protuberantes   **10.**
10. Plantas ca. 1–2 cm de altura; ca. 18 pares de hojas o
    más, hojas 2–2.5 mm de longitud   *F. asplenioides*
10. Plantas ca. 0.5 cm de altura; 3–10 pares de hojas,
    raramente más, hojas 1–2 mm de longitud.
                                   F. prionoides

Fissidens asplenioides Hedw., Spec. Musc. 156. 1801. (Fig. 20A).
Pastaza: Río Bobonaza, 460 m, *Spruce 506, 507* (NY; Mitten, 1869). Illustration based on *Fransén 72* (AAU) from Ecuador.
In Ecuador reported also from Pichincha and Tungurahua; at elevations from 460–2770 m.

Fissidens diplodus Mitt., J. Linn. Soc., Bot. 12: 589. 1869.
Synonym: *Fissidens muriculatus* Spruce in Mitt.
Pastaza: Río Bobonaza, 365 m, *Spruce 474, 475* (NY; Mitten, 1869).

Fissidens flavinervis Mitt.
Sucumbíos: Reserva Faunística Cuyabeno, near Laguna Grande, 00°00', 76°12'W, 265 m, *Balslev 84925* (AAU, NY, PAC). Determined by R. Pursell.

Fissidens hydropogon Spruce in Mitt., J. Linn. Soc., Bot. 12: 585. 1869.
Pastaza: Río Bobonaza, 460 m, *Spruce 506, 507* (NY; Mitten, 1869).
This species, the only aquatic *Fissidens* found on rocks and branches in streams in Amazonas, has recently been reviewed by Pursell *et al.* (1988), and is still known only from the type locality along Río Bobonaza where it was collected by Spruce.

Fissidens inaequalis Mitt., J. Linn. Soc., Bot. 12: 589. 1869. (Fig. 20B).
Napo: Añangu, ca. 75 km east of Coca, ca. 00°32'S, 76°23'W, 245–325 m, *Churchill & Sastre-De Jesús 13858a* (NY; Churchill *et al.*, 1992). Illustration based on *Churchill & Sastre-De Jesús 13858a.*

Fissidens intermedius C. Müll., Linnaea 21: 181. 1848. (Fig. 19C).
Napo: Añangu, ca. 75 km east of Coca, ca. 00°32'S, 76°23'W, 245–325 m, *Churchill & Sastre-De Jesús 13858b* (NY; Churchill *et al.*, 1992). Illustration based on *Churchill & Sastre-De Jesús 13858b.*

Fissidens intramarginatus (Hampe) Mitt., J. Linn. Soc., Bot. 12: 242. 1869.
Napo: Añangu, ca. 75 km east of Coca, ca. 00°32'S, 76°23'W, 245–325 m, *Churchill & Sastre-De Jesús 13872* (NY; Churchill *et al.*, 1992). Illustration based on *Churchill & Sastre-De Jesús 13872.*

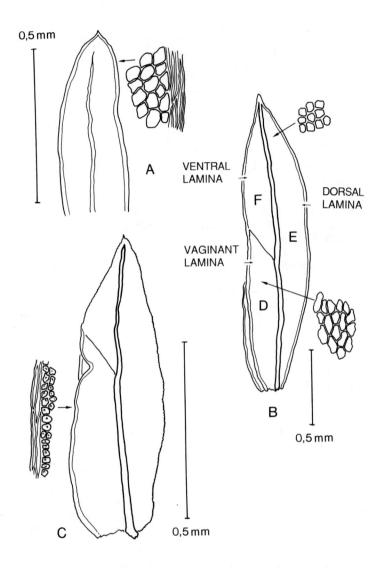

Figure 19. A. *Fissidens mollis*, A. upper half of leaf; B. *Fissidens zollingeri*, B. leaf illustrating terminology - D = vaginant lamina, E = dorsal lamina, F = ventral lamina; C. *Fissidens intermedius*, C. leaf.

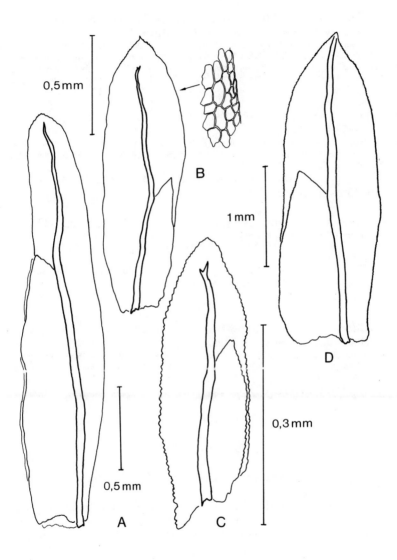

Figure 20. A. *Fissidens asplenioides*, A. leaf; B. *Fissidens inaequalis*, B. leaf; C. *Fissidens steerei*, C. leaf; D. *Fissidens prionodes*, D. leaf.

Fissidens mollis Mitt., J. Linn. Soc., Bot. 12: 600. 1869. (Fig. 19A).
Synonym: *Fissidens macrophyllus* Mitt.

Napo: Añangu, ca. 75 km east of Coca, ca. 76°23'W, 00°32'S, 245–325 m, *Churchill &
Sastre-De Jesús 13813-a* (AAU, NY; Churchill *et al.*, 1992); Ca. 24 km S of Tena on Tena–Puyo
road, 77°51'W, 01°12'S, ca. 600 m, *Churchill & Sastre-De Jesús 13667* (NY). **Pastaza:** Río
Bobonaza, 395 m, *Spruce 532* (NY; Mitten, 1869). Illustration based on *Churchill & Sastre-De
Jesús 13813-a.*

Fissidens prionodes Mont., Ann. Sc. Nat. Bot. sér. 2, 3: 200. 3 *f. 1*.
1835. (Fig. 20D).

Napo: Añangu, ca. 75 km east of Coca, ca. 00°32'S, 76°23'W, 245–325 m, *Churchill &
Sastre-De Jesús 13810* (NY; Churchill *et al.*, 1992). **Pastaza:** Río Bobonaza, 365 m, *Spruce 495*
(NY; Mitten, 1869).

In Ecuador reported also from the province of Azuay; at
elevations from 245–2500 m.

Fissidens steerei Grout, N. Amer. Fl. 15 (3): 191. 1943. (Fig. 20C).

Napo: Añangu, ca. 75 km east of Coca, ca. 00°32'S, 76°23'W, 245–325 m, *Churchill &
Sastre-De Jesús 13813-b* (AAU, NY; Churchill *et al.*, 1992). Illustration based on *Churchill &
Sastre-De Jesús 13813-b.*

This is a very small species, 3–5 mm tall, and leaves ca. 0.3–0.5
mm long. Known also from Mexico, Guatemala, Venezuela and
Colombia.

Fissidens zollingeri Mont., Ann.Sc. Nat. Bot. ser. 3, 4: 114. 1845.
(Fig. 19B). Synonym: *Fissidens kegelianus* C. Müll.

Napo: Añangu, ca. 75 km east of Coca, ca. 00°32'S, 76°23'W, 245–325 m, *Churchill &
Sastre-De Jesús 13811, 13815* (NY; Churchill *et al.*, 1992). Illustration based on *Churchill &
Sastre-De Jesús 13811.*

At elevations from 25-325 m.

HYDROPOGONACEAE

Hydropogon Brid.

Hydropogon fontinaloides (Hook.) Brid., Bryol. Univ. 1: 770. 1826.
(Fig. 38B-D).

Plants usually dull green, forming mats; primary stems creeping,
secondary stems spreading or more often pendent, often devoid of
leaves or leaves often distant, more foliate on distal branches and
about inflorescence, leaves oblong to oblong-obovate, 1.5–2.2 mm
long, 0.7–1.0 mm wide, short acuminate, margins recurved, plane
at apex, serrulate to serrate in distal 1/2–2/3 of lamina, costae
absent or short, forked and weak, median cells oblong-fusiform to
fusiform rhomboidal, smooth, alar cells rounded-quadrate to
-rectangular, not inflated; perichaetial leaves long oblong-

lanceolate; setae very short, ca. 0.1 mm long, capsules immersed, erect, urn ovoid, 0.6–0.8 mm long, peristome teeth 16, exostome striate, opercula conic-mamillate, 0.2–0.3 mm long; calyptrae not observed.

Our collections have been found as epiphytes on trees, usually associated with inundated sites, *i.e.*, along rivers. Distributed in the upper Amazon basin and extending to French Guiana, Suriname and Venezuela. References: Churchill (1991a), Welch (1943).

Napo: Río Yasuní, 3–4 km from Río Napo, 00°57'S, 75°25'W, 260 m, *Holm-Nielsen et al., 19840* (AAU); near Tena, 6 km along Río Pano, 00°58'S, 77°52'W, 600 m, *Holm-Nielsen & Jeppesen 684* (AAU). Sucumbíos: Reserva Faunística Cuyabeno, near Laguna Grande, 00°00', 76°12'W, 265 m, *Balslev 84920*, (AAU); Río Cuyabeno, near Puerto Montúfar, 00°06'S, 76°01'W, 230 m, *Holm-Nielsen et al. 21239* (AAU). Illustration based on *Holm-Nielsen et al. 19840*.

HYPNACEAE

Plants forming mats; stems mostly spreading, occasionally subascending, leaves erect or falcate, similar or dimorphic, or branch leaves differentiated from stem leaves, costae mostly short (1/5 or less lamina length) and forked, or absent, median cells elongate, alar region differentiated, cells few, mostly quadrate or short rectangular, or not differentiated; sporophyte lateral, setae elongate, smooth to rarely distally roughened, capsules horizontal to pendulous, urn mostly short ovoid, peristome double, exostome teeth 16, striate below, often papillose distally, endostome membrane usually high, segments 16, cilia 1–2; opercula conic to rostrate; calyptrae cucullate.

Species of this family are often found in shaded places on logs, occasionally trunks of trees and shrubs, and also on soil.

Key to the Genera

1. Median leaf cells smooth   **2.**
   2. Cells broadly fusiform to hexagonal, lax   *Vesicularia*
   2. Cells linear, not lax   **3.**
      3. Leaves dimorphic, upper median and lateral leaves ovate-lanceolate, underside leaves small, narrowly lanceolate
           Rhacopilopsis
      3. Leaves similar though median and lateral leaves often differentiated   **4.**
         4. Leaves falcate-secund; plants glossy-green
             Ectropothecium
         4. Leaves erect-spreading, not falcate; plants pale green
             Isopterygium
1. Median leaf cells papillose, papillae over cell lumen or projecting at angles   **5.**

5. Leaves ovate-oblong, apex truncate; papillae over cell lumen
Phyllodon
5. Leaves ovate-lanceolate, apex acuminate to acute; papillae projecting at cell angles **6.**
  6. Stem leaves ±complanate, lateral leaves oblong-lanceolate, folded *Taxiphyllum*
  6. Stem leaves not complanate, leaves ovate-lanceolate, not folded **7.**
    7. Plants often arching, stems stipate; secondary stem leaves and branch leaves dimorphic, stem or stipe leaves abruptly and narrowly long acuminate
Mittenothamnium
    7. Plants procumbent, stems not stipate; secondary stem leaves similar, not strongly differenitated, stem leaves ovate- lanceolate, gradually acuminate
Chryso-hypnum

Clave para los Géneros

1. Células medias de la hoja lisas **2.**
  2. Células ampliamente fusiformes hasta hexagonales, laxas
Vesicularia
  2. Células lineares, no laxas **3.**
    3. Hojas dimórficas, hojas laterales y medias superiores ovado-lanceoladas, hojas del lado inferior pequeñas, estrechamente lanceoladas *Rhacopilopsis*
    3. Hojas similares, hojas medias y laterales a menudo diferenciadas **4.**
      4. Hojas falcadas-secundiformes; plantas verde brillantes
Ectropothecium
      4. Hojas erectas-postradas, no falcadas; plantas verde-blancuzcas *Isopterygium*
1. Células medias de la hoja papilosas, papilas sobre el lumen celular o proyectandose hacia los ángulos **5.**
  5. Hojas ovado-oblongas, ápice truncado; papilas sobre el lumen celular *Phyllodon*
  5. Hojas ovado-lanceoladas, ápice acuminado hasta agudo; papilas proyectandose hacia los ángulos de la célula **6.**
    6. Hojas del tallo ±complanadas, hojas laterales oblongo-lanceoladas, plegadas *Taxiphyllum*
    6. Hojas del tallo no complanadas, hojas ovado-lanceoladas, no plegadas **7.**
      7. Plantas a menudo arqueadas; tallos estipitados; hojas del tallo secundario y ramas dimórficas, hojas del tallo o pecíolo abruptamente estrechas y largamente acuminadas *Mittenothamnium*
      7. Plantas procumbentes; tallos no estipitados; hojas del tallo secundario similares, no fuertemente

diferenciadas, hojas del tallo ovado-lanceoladas,
gradualmente acuminadas *Chryso-hypnum*

Chryso-hypnum Hampe

Chryso-hypnum diminutivum (Hampe) Buck, Brittonia 36: 182.
1984. (Fig. 21A-B). Synonyms: *Mittenothamnium diminutivum*
(Hampe) Britt., *M. subthelistegum* Card., *M. thelistegum* (C.
Müll.) Card.

Plants procumbent, forming mats; stems and branches
spreading, leaves ovate-lanceolate, ca. 1 mm or less long, similar to
branch leaves, margins recurved below, serrulate in distal 2/3 or
more, costae short and forked, ca. 1/4 lamina length, median cells
oblong-linear, both lower and upper angles projecting; setae 25–30
mm long, wiry, capsule short, ca. 1 mm long, curved, opercula
conic-short rostrate.

A single species of this genus is recorded for Ecuador. *Chryso-
hypnum* has only recently been segregated from *Mittenothamnium*,
largely based on the characters given in the generic keys.

Napo: El Napo, 500 m, *Benoist 4679* (Thériot, 1936). **Pastaza:** Río Bobonaza, 460 m,
Spruce 1087 (NY; Mitten, 1869). Illustration based on *Andersson 761* (AAU) from Ecuador.

In Ecuador reported also from Morona-Santiago; at elevations
from 460–1000 m.

Ectropothecium Mitt.

Plants small to medium sized; stems spreading, leaves falcate-
secund, narrowly ovate-lanceolate to lanceolate-subulate, apex
acuminate, base undifferentiated, margins distally serrulate to
serrate, costae absent or indistinct, short and forked, median cells
linear, smooth, alar region not or weakly differentiated, basal and
insertion cells rectangular; autoicous; capsules inclined or
pendulous, urn ovoid; opercula mammillose or short rostrate.

Only the two species listed below are known from Ecuador, both
typically are found on logs, less common on ground litter or tree or
treelet trunks.

Key to the Species

1. Leaves ovate-lanceolate *E. aeruginosum*
1. Leaves narrow, lanceolate-subulate *E. leptochaeton*

Clave para las Especies

1. Hojas ovado-lanceoladas *E. aeruginosum*
1. Hojas estrechas, lanceolado-subuladas *E. leptochaeton*

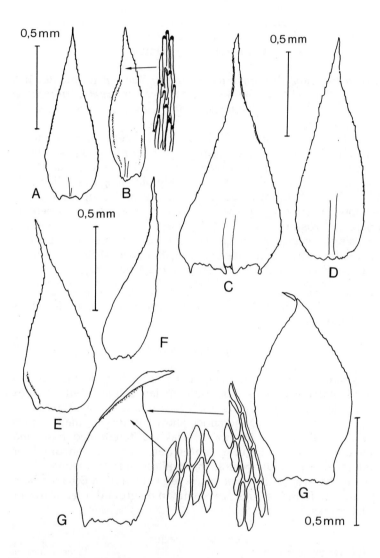

Figure 21. A-B. *Chryso-hypnum diminutivum*, A. stem leaf, B. branch leaf; C-D. *Mittenothamnium reptans*, C. stem leaf, D. branch leaf; E-F. *Rhacopilopsis trinitensis*, E. lateral leaf, F. underside (ventral) leaf; G. *Vesicularia vesicularis*, G. stem leaves.

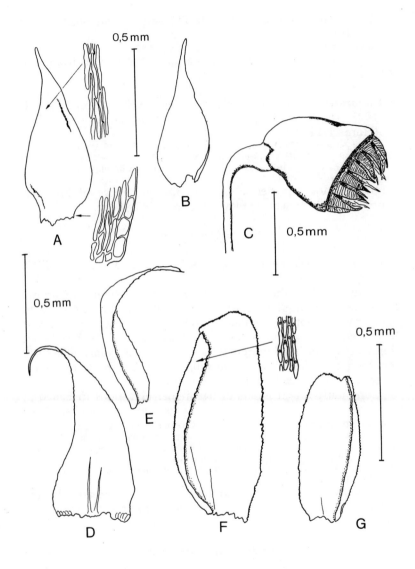

Figure 22. A-C. *Isopterygium tenerum*, A. stem leaf, B. branch leaf, C. capsule; D-E. *Ectropothecium leptochaeton*, D. stem leaf, E. branch leaf; F-G. *Phyllodon truncatulus*, F. stem leaf, G. branch leaf.

Ectropothecium aeruginosum (C. Müll.) Mitt., J. Linn. Soc., Bot. 12: 513. 1869.

Morona-Santiago: Yurupaz, 600 m, *Harling 2261* (Crum, 1957).

In Ecuador reported also from Pastaza; at elevations from 600–925 m.

Ectropothecium leptochaeton (Schwaegr.) Buck, Brittonia 35: 311. 1983. (Fig. 22D-E). Synonym: *Ectropothecium apiculatum* (Hornsch.) Mitt.

Napo: Añangu, ca. 75 km east of Coca, ca. 00°32'S, 76°23'W, 245–325 m, *Churchill & Sastre-De Jesús 13793-b, 13857, 13862b* (AAU, NY; Churchill *et al.*, 1992). **Pastaza:** Finca Valle de Muerte on Río Curaray, ca. 200 m, *Andersson 901b* (GB). Illustration based on *Churchill & Sastre-De Jesús 13793-b.*

In Ecuador reported also from the provinces of Morona-Santiago and Pichincha; at elevations from 245–900 m.

Isopterygium Mitt.

Isopterygium tenerum (Sw.) Mitt., J. Linn. Soc., Bot. 12: 499. 1869. (Fig. 22A-C).

Plants small, rather delicate, forming mats; stems spreading, leaves loosely complanate, ovate-lanceolate, 0.7–1.0 mm long, to 0.3 mm wide, ±long acuminate, margins plane or slightly recurved below, entire, ecostate, median cells linear, smooth, alar cells not differentiated or basal margin cells 2–4, subrectangular; setae to 10 mm long; capsules horizontal to subpendent, urn short-oblong, ca. 0.8 mm long; opercula short-rostrate.

Commonly found in somewhat higher elevations in Ecuador, *Isopterygium tenerum* is found on logs and tree trunks. Three species are recorded for Ecuador. Reference: Ireland (1991).

Napo: Añangu, ca. 75 km east of Coca, ca. 00°32'S, 76°23'W, 245–325 m, *Churchill & Sastre-De Jesús 13862-a* (AAU; Churchill *et al.*, 1992). **Pastaza:** Río Bobonaza, 395 m, *Spruce 1064* (NY; Mitten, 1869). **Sucumbíos:** Reserva Faunística Cuyabeno, N of Laguna Grande, 00°01'N, 76°11'W, 265 m, *Heikkinen RH-1990-333* (NY). Illustration based on *Andersson 944* (AAU) from Ecuador.

In Ecuador reported from Morona-Santiago, Napo and also the Galapagos Islands; at elevations from 245 m to probably more than 1000 m.

Mittenothamnium Henn.

Mittenothamnium reptans (Hedw.) Card., Rev. Bryol. 40: 21. 1913. (Fig. 21C-D).

Plants forming loose to dense mats; stipitate secondary stems, spreading or ascending, leaves broadly ovate-lanceolate or triangular, ca. 1.2 mm long, to 0.5 mm wide, apex abruptly long acuminate, base often ±decurrent, margins serrate, costae short and forked, median cells linear, smooth to more often distinctly papillose at upper cells by projecting angles; branch leaves differentiated from stem leaves, small, lanceolate to ±triangular, to 1.4 mm long, median cells often strongly papillose by projecting angles; setae elongate; capsules inclined; operula rostrate.

A highly variable and poorly known genus with possibly only a few species in Ecuador, certainly not the 11 species presently recorded.

Pastaza: Río Bobonaza, 460 m, *Spruce 1117* (NY; Mitten, 1869). Illustration based on *Churchill et al. 13465* (NY) from Colombia.

In Ecuador reported also from Chimborazo, Morona-Santiago, Pichincha, and Tungurahua; at elevations from 460–2800 m.

Phyllodon B.S.G.

Phyllodon truncatulus (C. Müll.) Buck, Mem. N. Y. Bot. Gard. 45: 521. 1987. (Fig. 22F-G). Synonym: *Glossadelphus truncatulus* (C. Müll.) Fleisch.

Plants green to yellowish-green, forming thin mats; stems spreading, leaves complanate, differentiated, lateral leaves short ovate-oblong, 0.6–0.9 mm long, 0.3–0.4 mm wide, concave, apex obtuse to truncate, margins serrulate throughout, costae double, short, ca. 1/5–1/4 lamina length, median cells oblong-linear, papillose, papillae projecting and usually over cell lumen; dioicous; setae smooth to distally roughened, capsules curved, inclined, urn ovoid; opercula conic.

A small pantropical genus, with one species presently recognized for Ecuador. *Phyllodon truncatulus* is rather widespread in the lowlands of the Neotropics. Reference: Buck (1987b).

Napo: Añangu, ca. 75 km east of Coca, ca. 00°32'S, 76°23'W, 245–325 m, *Churchill & Sastre-De Jesús 13820* (AAU, NY; Churchill *et al.*, 1992). **Pastaza:** Río Bobonaza, *Spruce s.n.* (NY; Mitten, 1869). Illustration based on *Churchill & Sastre-De Jesús 13820.*

Rhacopilopsis Ren. & Card.

Rhacopilopsis trinitensis (C. Müll.) Britt. & Dix., J. Bot. 60: 86, 88 *1 f. 4–5.* 1927. (Fig. 21E-F).

Plants small, mostly golden-yellow or brown; stems spreading, leaves complanate, dimorphic, upper median and lateral leaves asymmetric, ovate-lanceolate, 1–1.3 mm long, 0.3–0.4 mm wide, abruptly long acuminate, margins plane, 1/3–2/3 serrulate distally, costae short and forked or absent, median cells oblong-linear, smooth, alar cells subquadrate to rectangular, dark golden-brown, underside leaves small, narrowly lanceolate.

A single widespread species found both in the Neotropics and Africa. *Rhacopilopsis* is rather inconspicuous and readily overlooked, typically found on logs and tree trunks.

Napo: Near Tena, 6 km along Río Pano, 600 m, *Holm-Nielsen & Jeppesen 702 p.p.* -with *Octoblepharum cocuiense* (AAU). **Pastaza:** Curaray SE of airstrip, 01°22'S, 76°57'W, 250 m, *Holm-Nielsen et al. 22281* (AAU); Lorocachi, along Río Curaray, 01°38'S, 75°58'W, 200 m, *Jaramillo et al. 31175 p.p.* - with *Syrrhopodon leprieurii* (AAU). **Sucumbíos:** Reserva Faunística Cuyabeno, N of Laguna Grande, 00°01'N, 76°11'W, 265 m, *Heikkinen RH-1990-155* (NY). Illustration based on *Holm-Nielsen et al. 22281*.

Taxiphyllum Fleisch.

Taxiphyllum taxirameum (Mitt.) Fleisch., Musci Fl. Buitenzorg 4: 1435. 1923. Synonym: *Taxiphyllum planissimum* (Mitt.) Broth.

Plants yellowish-green, forming mats; stems spreading, leaves ±complanate, lateral leaves ±asymmetric, oblong-lanceolate, ca. 1.5 mm long, acuminate, margin usually folded on one side, plane distally, recurved at base, serrulate throughout, costae weakly short and forked, median cells fusiform-linear, projecting as cell angles, alar cells little differentiated, some subquadrate; dioicous.

Pastaza: Río Bobonaza, 395 m, *Spruce 1061* (NY; Mitten, 1869).

At elevations from 395–1000 m. Two species of *Taxiphyllum* are recorded for Ecuador.

Vesicularia (C. Müll.) C. Müll.

Vesicularia vesicularis (Schwaegr.) Broth., Nat. Pfl. 1(3): 1094. 1908. (Fig. 21G). Synonym: *Vesicularia amphibola* (Mitt.) Broth.

Plants small, glossy, forming mats; stems spreading, leaves crispate and homomallous when dry, secund, complanate when wet, lateral leaves ovate-lanceolate to broadly ovate, asymmetric, 0.8–1.1 mm long, 0.4–0.6 mm wide, abruptly short acuminate, margins plane, entire to bluntly serrulate distally, ecostate, occasionally short and forked, median cells fusiform, rhomboidal to hexagonal, somewhat lax, distal margin cells narrow, median leaves symmetric, ovate-lanceolate, smaller than lateral leaves; autoicous; setae to 18 mm long; capsules pendulous, urn short cylindrical, ca. 0.8 mm long; opercula short rostrate, oblique.

Vesicularia is a small genus in the Neotropics, two species are recorded for Ecuador. The genus is commonly found on logs, ground litter, and less often on soil.

Morona-Santiago: Patuca, 600 m, *Harling 2275* (S), *2276* (S), *2278b* p.p., *2279* (S), *2282, 2286b* (Crum, 1957). **Napo:** El Napo, 500 m, *Benoist 4689* (Thériot, 1936); Las Sachas, on Coca–Lago Agrio road, km 40, ca. 250 m, *Fransén 53, 61* (AAU); Lago Agrio–El Conejo road, ca. 300 m, *Andersson 736* (AAU); Río Yasuní, 3–4 km from Río Napo, 00°57'S, 75°25'W, 260 m, *Holm-Nielsen et al. 19881* (AAU); Río Aguarico, San Pablo de los Secoyas, 00°17'S, 76°26'W, 235 m, *Holm-Nielsen et al. 21071* (AAU); At Río Payamino, 60 km along Río Payamino, W of Coca, 00°29'S, 77°12'W, 350 m, *Holm-Nielsen & Jeppesen 762* (AAU); Añangu, ca. 75 km east of Coca, ca. 00°32'S, 76°23'W, 245–325 m, *Churchill & Sastre-De Jesús 13742b, 13778, 13797-a, 13826* (NY; Churchill *et al.*, 1992); Río Suno, 00°42'S, 77°10'W, 400 m, *Holm-Nielsen & Jeppesen 889, 891, 902,* (AAU); Archidona, 00°54'S, 77°48'W, 600 m, *Holm-Nielsen & Jeppesen 1055* (AAU); Yuralpa, 00°55'S, 77°21'W, 440 m, *Holm-Nielsen & Jeppesen 941* (AAU, GB); Robinson *et al.*, 1971); Río Yasuní, Charapillo, 01°03'S, 75°44'W, 320 m, *Holm-Nielsen et al. 19957* (AAU). **Pastaza:** Near Puyo, ca. 400 m, *Mexía 6956* (GB); Namuyacu, tributary of Río Curaray, ca. 200 m, *Andersson 894* (GB); Curaray, ca. 200 m, *Andersson 867* (AAU, GB); Curaray, northern bank 2 km W of school, 01°22'S, 76°58'W, 250 m, *Holm-Nielsen et al. 21936* (AAU); Ceilán, Pica from Ceilán to Río Cononaco along Río Curaray, 01°36'S, 75°40'W, 200 m, *Brandbyge & Asanza C. 31673* (AAU); Montalvo, 0.1 km north and east of military camp, 02°05'S, 76°58'W, ca. 250 m, *Løjtnant & Molau 13509* (AAU), *13383* (GB); Río Pastaza, between Destacamento Chiriboga and Apachi Entza, ca. 02°20–32'S, 76°55'W, ca. 285 m, *Øllgaard et al. 35188* (AAU). **Sucumbíos:** Reserva Faunística Cuyabeno, near Laguna Grande, 00°00', 76°12'W, 265 m, *Balslev 84922* (AAU); *Heikkinen RH-1990-85c & -370a* (NY). Illustration based on *Balslev 84922*.

In Ecuador reported also from the provinces of Pastaza, Pichincha and Tungurahua; at elevations from 245–2155 m.

The majority of our collections fit the concept of *Vesicularia amphibola* which exhibits a rather narrow border, other collections agree with the concept of *V. vesicularis* which lacks a distinct border. However, some intermediate collections were observed between both, thus I have at present recognized but a single species for Amazonas. *Vesicularia* can be easily confused with *Leucomium*, for differences between the two species see comments under *Leucomium*.

LEPTODONTACEAE

Pseudocryphaea Brid. ex Broth.

Pseudocryphaea domingensis (Spreng.) Buck, Bryologist 83: 455. 1980. (Fig. 23A-C). Synonym: *Pseudocryphaea flagellifera* (Brid.) Britt.

Plants stiffly dendroid; primary stems creeping, secondary stems ascending, distal branches with microphyllous branches from leaf axils; leaves ovate, 1.2–2.0 mm long, 0.6–1 mm wide, short acuminate, margins plane, entire to serrulate distally, costae single, ±weak, ca. 2/3–3/4 lamina length, median cells fusiform, smooth to

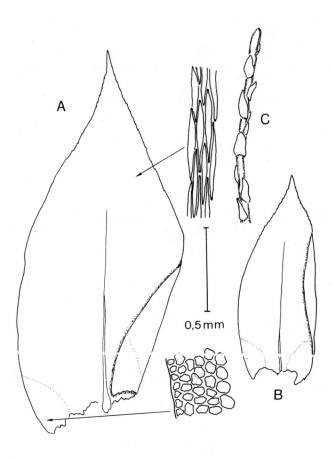

Figure 23. A-C. *Pseudocryphaea domingensis,* A. stem leaf, B. branch leaf, C. microphyllous branch.

indistinctly unipapillose by projecting cell angles, alar region distinct, cells subquadrate or rounded, ±sinuose; microphyllous branches slender, leaves ovate-lanceolate, to 0.2 mm long, ecostate; seta elongate; capsules ovoid (sporophytes not observed in Ecuador).

Only recently reported for Ecuador, this single widely distributed neotropical species is found in wet lowland and premontane forests; in shade or semi-exposed places, on trunks and branches of trees, treelets, and lianas.

Napo: Añangu, ca. 75 km east of Coca, ca. 00°32'S, 76°23'W, 245–325 m, *Churchill & Sastre-De Jesús 13847, 13865* (NY; Churchill *et al.*, 1992). Illustration based on *Churchill et al. 14834* (NY) from Colombia.

LEUCOBRYACEAE

Plants white or whitish-green, occasionally tinged with purple or red, erect, forming loose to dense tufts or cushions; leaves often with expanded oval to oblong base, ligulate to linear-lanceolate above, costae appearing absent or single (*Leucophanes*), in cross-section composed of leucocysts (hyaline cells) above and below internal layer of smaller chlorocysts (chlorophyll cells), cells smooth; setae erect, ±short or elongate, smooth, capsules erect or curved, urn ovoid-cylindrical, peristome single, teeth 8 or 16, often divided; opercula rostrate; calyptrae cucullate, smooth.

In Ecuador this family is found from wet lowland to montane forests mostly as epiphytes or on logs.

Key to the Genera
1. Leaves with a median costa of a fascicle of stereids close to back surface *Leucophanes*
1. Leaves lacking a median costa near back surface **2.**
   2. Leaf in cross-section with chlorocysts 4-angled; upper leaf surface concave; capsules inclined, asymmetric *Leucobryum*
   2. Leaf in cross-section with chlorocysts 3-angled; upper leaf surface flat; capsules erect, symmetric *Octoblepharum*

Clave para los Géneros
1. Hojas con costa media de un fascículo de estereidas cerca a la superficie inferior *Lecophanes*
1. Hojas sin costa media cerca a la superficie inferior **2.**
   2. Hojas en sección transversal con clorocistes 4-angulados; superficie superior de la hoja cóncava; cápsulas inclinadas, asimétricas. *Leucobryum*

2.   Hojas en sección transversal con clorocistes 3-angulados;
superficie superior de la hoja plana; cápsulas erectas, simétricas
Octoblepharum

Leucobryum Hampe

Leucobryum martianum (Hornsch.) Hampe, Linnaea 17: 317. 1843.
(Fig. 24A-C).
Plants with leaves often falcate-secund, ovate-lanceolate, 4–6
mm long, 0.5–1.0 mm wide, concave below, becoming narrow and
channeled above, in cross-section one row of leucocysts above and
below chlorocyst row, distally chlorocysts closer to dorsal surface,
chlorocysts 4-sided; setae 2–2.5 cm long; capsules inclined,
asymmetric, strumose, furrowed when dry, peristome teeth 16,
divided ca. 2/3, striate below.
Five species of *Leucobryum* are known from Ecuador, mostly
from higher elevations.
Pastaza: Lorocachi, along Río Curaray, below military camp, 01°38'S, 75°58'W, 200 m,
Jaramillo et al. 31389 (AAU); Río Bobonaza, between Destacamento Cabo Pozo and La Boca,
ca. 2°30–35'S, 76°38'W, ca. 275 m, *Øllgaard et al. 34942* (AAU). **Sucumbíos:** Reserva Faunística
Cuyabeno, N of Laguna Grande, 00°01'N, 76°11'W, 265 m, *Heikkinen RH-1990-160* (NY).
Illustration based on *Øllgaard et al. 34942*.
In Ecuador reported also from the province of Morona-
Santiago; at elevations from 200–2000 m.

Leucophanes Brid.

Leucophanes molleri C. Müll., Flora 69: 285. 1886. (Fig. 24D-G).
Synonyms: *Leucophanes calymperatum* C. Müll., *L. mittenii*
Card. *in* Par.
Plants ±small, delicate and somewhat fragile, to 1.5 cm tall.
Leaves narrowly lanceolate to linear-lanceolate, 3–4 mm long, to
0.3 mm wide, distally ±channeled, apex apiculate to rather blunt,
margins plane to slightly recurved, bordered with few linear cells,
serrulate in distal 1/4 of lamina, median costa readily observed, in
cross-section costa composed of a fasciculate bundle of stereids
close to the back surface, remaining lamina in cross-section with
one layer of leucocysts above and below triangular-shaped
chlorocysts; propagula often present on leaf apices. Sporophytes
not observed.
Only a single species of this genus is known from Ecuador;
found on trunks and branches of trees or treelets. Reference:
Salazar Allen (1993).

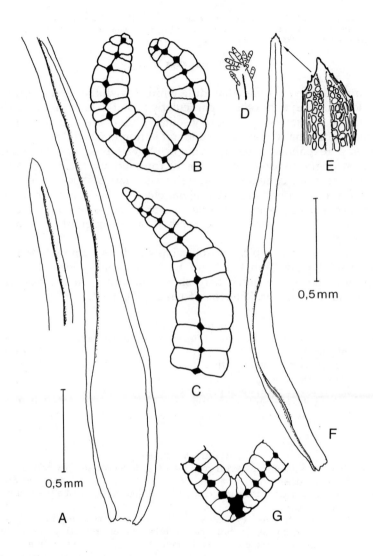

Figure 24. A-C. *Leucobryum martianum*, A. leaf, B. distal leaf x-section, C. portion of leaf base x-section; D-G. *Leucophanes molleri*, D. distal leaf tip with propagulae, E. detail of leaf apex, F. leaf, G. portion of leaf base x-section.

92 S. P. Churchill

Napo: Las Sachas, Coca-Lago Agrio road km 40, ca. 250 m, *Fransén 52A* (AAU, GB).
Sucumbíos: Reserva Faunística Cuyabeno, N of Laguna Grande, 00°01'N, 76°11'W, 265 m,
Heikkinen RH-1990-133 (NY). Illustration based on *Fransén 52A.*

Octoblepharum Hedw.

Plants glossy to somewhat dull white to whitish-green, often
reddish or purplish colored, particularly at leaf base, forming loose
to dense cushions or tufts, rarely solitary; leaves erect-spreading to
erect, occasionally ±falcate-secund, often fragile when dry, ligulate
from a ±expanded suboval, ovate or obovate base, distally ±flat,
not strongly concave, apex mostly rounded-apiculate, margins
entire to more commonly serrate or serrulate at apex, in cross-
section costae consisting of triangular chlorocysts with one or more
layers of leucocysts above and below; autoicous; setae elongate, urn
cylindrical, peristome single, teeth 8 or 16; opercula rostrate,
oblique.

Octoblepharum, typical of tropical lowlands, is commonly found
in shaded placed, on trunks of trees and lianas. Several more
species than those given here should be found in the eastern
lowlands of Ecuador. Reference: Salazar Allen (1991).

Key to the Species
1. Leaf in cross-section with 1 layer of leucocysts above and below
   chlorocysts   *O. cocuiense*
1. Leaf in cross-section with 2 rows of leucocysts above and below
   chlorocysts   **2.**
   2. Leaves fragile, distal tips often broken off; upper cells of expanded
      leaf base irregularely isodiametric   *O. pulvinatum*
   2. Leaves not fragile; upper cells of expanded leaf base rectangular
          O. albidum

Clave para las Especies
1. Hoja en sección transversal con una capa de leucocistes arriba y abajo
   de los clorocistes   *O. cocuiense*
1. Hoja en sección transversal con 2 filas de leucocistes arriba y abajo de
   los clorocistes   **2.**
   2. Hojas frágiles, extremos distales a menudo quebradizos; base
      extendida de la hoja con células superiores irregularmente
      isodiamétricas; perístoma dentado   *O. pulvinatum*
   2. Hojas no frágiles, base extendida de la hoja con células superiores
      rectangulares; perístoma con ocho dientes   *O. albidum*

Octoblepharum albidum Hedw., Spec. Musc. 50. 1801. (Fig. 25A)
    Napo: Añangu, ca. 75 km east of Coca, ca. 00°32'S, 76°23'W, 245–325 m, *Churchill &*
Sastre-De Jesús 13815 (NY; Churchill et al., 1992); Río Lagarto Cocha, near Redondo Cocha &

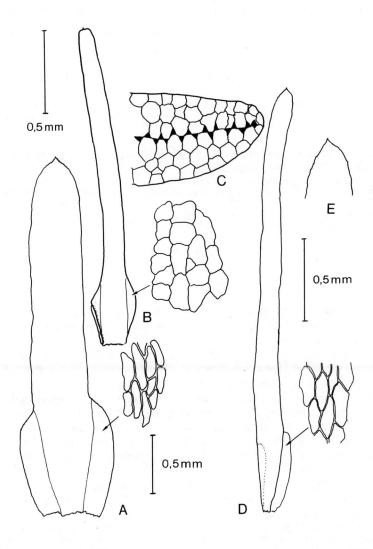

Figure 25. A. *Octoblepharum albidum*, A. leaf; B-C. *Octoblepharum pulvinatum*, B. leaf, C. x-section at mid leaf; D-E. *Octoblepharum cocuiense*, D. leaf, E. leaf apex.

Inauya Cocha, 00°35'S, 75°15'W, 190 m, *Lawesson et al. 44263* (AAU). **Pastaza:** Curaray, ridge NE of Destacamento, 01°21'S, 76°56'W, 250 m, *Holm-Nielsen et al. 22053* (AAU). **Sucumbíos:** Reserva Faunística Cuyabeno, near Laguna Grande, 00°00', 76°12'W, 265 m, *Balslev 84925* (AAU). Illustration based on *Balslev 84925.*

In Ecuador reported from Loja and Morona-Santiago, also the Galapagos Islands; at elevations from 245–1080 m.

Octoblepharum cocuiense Mitt., J. Linn. Soc., Bot. 12: 109. 1869. (Fig. 25D-E)

Napo: near Tena, 6 km along Río Pano, 00°58'S, 77°52'W, 600 m, *Holm-Nielsen & Jeppesen 680, 702* (AAU; Robinson et al., 1971). Illustration based on *Holm-Nielsen & Jeppesen 680.*

Octoblepharum pulvinatum (Dozy & Molk.) Mitt., J. Linn. Soc., Bot. 12: 109. 1869. (Fig. 25B-C)

Napo: Añangu, ca. 75 km east of Coca, ca. 00°32'S, 76°23'W, 245–325 m, *Churchill & Sastre-De Jesús 13815b, 13861* (AAU, NY; Churchill et al., 1992); Río Suno, 77°10'W, 00° 42'S, 400 m, *Holm-Nielsen & Jeppesen 906* (AAU; Robinson et al., 1971). **Sucumbíos:** Reserva Faunística Cuyabeno, N of Laguna Grande, 00°01'N, 76°11'W, 265 m, *Heikkinen RH-1990-162* (NY). Illustration based on *Churchill & Sastre-De Jesús 13815b.*

LEUCOMIACEAE

Leucomium Mitt.

Leucomium strumosum (Hornsch.) Mitt., J. Linn. Soc., Bot. 12: 502. 1869. (Fig. 26A). Synonyms: *Leucomium compressum* Mitt., *L. lignicola* Spruce ex Mitt.

Plants pale whitish- or yellowish-green, forming thin mats; stems spreading, loosely complanate; leaves crispate and contorted when dry, ovate-lanceolate, 1.8–2.4 mm long, 0.5–0.8 mm wide, long acuminate, margins plane, entire, ecostate, median cells fusiform to rhombodial, lax; setae elongate, to 14 mm long, smooth; capsules inclinded, urn ovoid, peristome double, exostome teeth 16, striate below, papillose above, furrowed, endostome basal membrane high, segments 16, cilia absent; opercula rostrate, oblique; calyptrae cucullate.

The only species for Ecuador, *Leucomium* is commonly found in shaded places on leaf litter, logs and base trunks of trees. Reference: Allen (1987).

Morona-Santiago: Patuca, 600 m, *Harling 2277 p.p.* (S; with *Callicostella pallida*). **Napo:** Añangu, ca. 75 km east of Coca, ca. 76°23'W, 00°32'S, 245–325 m, *Churchill & Sastre-De Jesús 13776* (NY; Churchill et al., 1992). **Sucumbíos:** Reserva Faunística Cuyabeno, N of Laguna Grande, 00°01'N, 76°11'W, 265 m, *Heikkinen RH-1990-168* (NY). Illustration based on *Churchill & Sastre-De Jesús 13643* (AAU) from Ecuador.

In Ecuador also reported Morona-Santiago; at elevations from 245–1200 m.

Vesicularia may be confused with *Leucomium* (Leucomiaceae), both are similar in habit and texture (when dry), and often have a similar leaf shape. *Vesiculeria* usually exhibits bluntly serrulate distal margins, and smaller cells, to 100 um long, whereas *Leucomium* exhibits entire margins and larger cells, to 180 um long.

METEORIACEAE

Plants forming mats or wefts, with primary stems creeping and secondary stems mostly pendent or pendulous, occasionally spreading (*Papillaria*), leaves concave or not, costae single or absent (*Pilotrichella*), median cells linear, smooth or papillose, alar cells often differentiated, basal cells often porose; setae short to ±elongate, capsules erect, urn ovoid, peristome double, exostome teeth 16, usually papillose, occasionally striate below, endostome basal membrane usually low, segments 16, cilia absent; opercula rostrate; calyptrae mitrate or cucullate.

Species of this family occurs commonly as epiphytes, less often on logs or soil banks. Eight genera are recorded for Ecuador, of which five are represented in Amazonas.

Key to the Genera

1. Costae absent; leaf apex acute-apiculate   *Pilotrichella*
1. Costae present; leaf apex long acuminate, piliferous or acute-apiculate
                                                                                            2.
    2. Median leaf cells pluripapillose, papillae over cell lumen, in a row
                                                                                     Papillaria
    2. Median leaf cells smooth   **3.**
        3. Alar group differentiated, forming a group of subquadrate thick-walled cells; leaves piliferous or broadly acute
                                                                               Squamidium
        3. Alar group little differentiated, not forming a distinct group of subquadrate cells; leaves long acuminate   **4.**
            4. Stem leaves spreading from the base, not clasping stem
                                                                          Meteoridium
            4. Stem leaves spreading or recurved distally, base clasping stem   *Zelometeorium*

Clave para los Géneros

1. Costas ausentes; ápice de la hoja agudo-apiculado   *Pilotrichella*
1. Costas presentes; ápice de la hoja largamente acuminado, pilífero o agudo-apiculado   **2.**

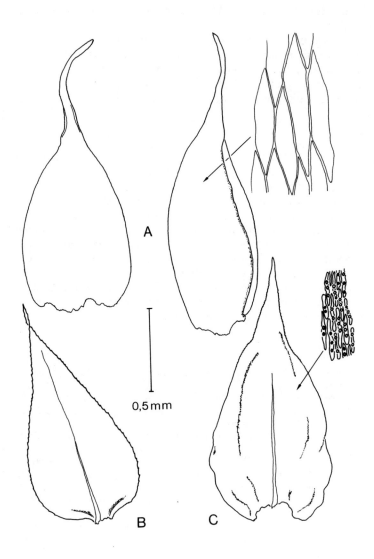

Figure 26. A. *Leucomium strumosum*, A. leaves; B. *Meteoridium remoti-folium*, B. leaf; C. *Papillaria nigrescens*, C. leaf.

2. Células medias de la hoja pluripapilosas, papilas sobre el lumen
   celular, en una fila   ***Papillaria***
2. Células medias de la hoja lisas   **3.**
    3. Grupo alar diferenciado, formando un grupo de células
       subcuadradas de paredes engrosadas; hojas pilíferas o
       ampliamente agudas   ***Squamidium***
    3. Grupo alar poco diferenciado; no forma un distintivo grupo
       de células subcuadradas; hojas largamente acuminadas   **4.**
        4. Hojas desde la base separadas del tallo, tallo no abrazado
                Meteoridium
        4. Hojas separadas del tallo o recurvadas distalmente,
           bases abrazando al tallo   ***Zelometeorium***

Meteoridium (C. Müll.) Manuel

Meteoridium remotifolium (C. Müll.) Manuel, Lindbergia 4: 49.
1977. (Fig. 26B). Synonym: *Meteoriopsis remotifolia* (C. Müll.)
Broth.

Secondary stems usually pendent, leaves spreading to wide-
spreading, ovate-lanceolate, 2.0–2.5 mm long, 1.0 mm wide, apex
long acuminate, base not clasping stem, not auriculate, margins
recurved at base, serrate distally, costae 1/2–2/3 lamina length,
median cells linear-vermicular, smooth; dioicous; setae short, ca.
equal to capusle length; urn ovoid, exostome teeth striate below;
opercula rostrate; calyptrae cucullate, naked.

More common in premontane to montane forests, this species,
the only representative of the genus, is widespread in the Neotropics.
Meteoridium may be confused with *Zelometeorium*, especially *Z.*
patulum, since the branch leaves are nearly identical, however as
noted in the keys the latter taxon exhibits stem leaves that are
clasping the leaf base; another feature of *Meteoridium* absent in
Zelometeorium are the distal young leaves that are cluster and
visiably sharp serrated margins. Reference: Manuel (1977a).

Napo: Añangu, ca. 75 km east of Coca, ca. 76°23'W, 00°32'S, 245–325 m, *Churchill &*
Sastre-De Jesús 13847 (NY; Churchill *et al.*, 1992). Illustration based on *Holm-Nielsen et al.*
4080 (AAU) from Ecuador.

In Ecuador reported also from Morona-Santiago; at elevations
from 245–1100 m.

Papillaria (C. Müll.) C. Müll.

Papillaria nigrescens (Hedw.) Jaeg., Ber. St. Gall. Naturw. Ges.
1875–1876: 265. 1877. (Fig. 26C).

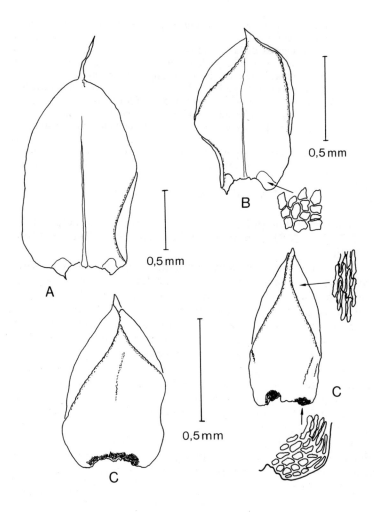

Figure 27. A. *Squamidium leucotrichum*, A. leaf; B. *Squamidium nigricans*, B. leaf; C. *Pilotrichella pentasticha*, C. leaves.

Plants forming mats or wefts, occasionally pendent, ±stiff; leaves ovate-lanceolate to broadly triangular, plicate, 1.5–2 mm long, 0.6–1 mm wide, apex acuminate, base auriculate, margins plane, serrulate, costae 2/3 lamina length, median cells linear-vermicular, pluripapillose, papillae in a row; dioicous; setae short; capsules slightly emergent, urn oblong, exostome papillose; calyptrae mitrate, hairy.

Papillaria, with four recorded species in Ecuador, is more common in the premontane and montane forests where is occurs as an epiphyte but also found on logs, soil and rocks.

Morona-Santiago: Along Médez–Sucúa road, 02°38'S, 78°13'W, 600 m, *Løjtnant & Molau 14566* (AAU). Illustration based on *Løjtnant & Molau 14566*.

In Ecuador reported from Azuay, Pichincha, Tungurahua, and also the Galapagos Islands; at elevations from ca. 600–2800 m.

Pilotrichella (C. Müll.) Besch.

Pilotrichella pentasticha (Brid.) Wijk & Marg., Taxon 9: 52. 1960. (Fig. 27C).

Plant secondary stems slender, usually pendent, secondary stem and branch leaves often appearing 5-ranked, ovate, ca. 1 mm long, concave, apex acute or acute-apiculate, often recurved, base auriculate, margins incurved and entire or serrulate distally, costa absent (rarely short and forked), median cells linear, ±vermicular, alar region poorly differentiated, cells quadrate to oval; setae elongate; urn ovoid, exostome papillose; opercula long rostrate; calyptrae cucullate.

Pilotrichella viridis (C. Müll.) Jaeg. reported from Morona-Santiago by Crum (1957) is likely a synonym of *P. pentasticha*. Typically epiphytic, this genus is represented by several species in the Neotropics. Five species are listed for Ecuador, however the only other common species, usually at higher elevations but possibly extending to the lowlands in Ecuador, is *P. flexilis* (Hedw.) Ångstr., distinguished by larger leaves (2–2.5 mm long), whereas *P. pentasticha* has smaller leaves (ca. 1 mm long).

Morona-Santiago: Yurupaz, 600 m, *Harling 2242b* (Crum, 1957). **Napo:** Shinguipino, between Río Napo and Tena, ca. 460 m, *Grubb et al. 2929* (Bartram, 1964); Río Payamino, 00° 29'S, 77°12'W, 350 m, *Holm-Nielsen & Jeppesen 760* (AAU, GB; Robinson et al., 1971). **Sucumbíos:** Río Güepi, 00°07–08'S, 75°30–39'W, ca. 200 m, *Brandbyge & Asanza C. 30610* (AAU). Illustration based on *Holm-Nielsen & Jeppesen 760*.

In Ecuador reported also from Chimborazo, Pastaza and Tungurahua; at elevations from 350–2155 m.

Squamidium (C. Müll.) Broth.

Plant secondary stems often pendent or spreading, glossy green, older stems and branches often golden or blackish; flagelliform branches often present; secondary stem leaves ovate to ovate-oblong, concave, apex piliferous or acute-apiculate, base decurrent, ±auriculate, margins distally serrulate, often incurved, costae 3/4 lamina length, weak, median cells linear, smooth, porose, alar region differentiated, cells subquadrate or oval, porose; dioicous; sporophytes immersed to exserted, setae equal or shorter than capsule; capsules ovoid-cylindrical, calyptrae mitrate, often hairy.

Five species of *Squamidium* are known from Ecuador; the genus is more commonly found at higher elevations in premontane and montane forests. *Orthostichopsis* may be confused with *Squamidium*, but that taxon typically has leaves in a distinct 5-spiral series. Reference: Allen and Crosby (1986b).

Key to the Species
1. Branch leaves ca. 2.5–4 mm long, piliferous   *S. leucotrichum*
1. Branch leaves ca. 0.8–1.5 mm long, acute or apiculate   *S. nigricans*

Clave para las Especies
1. Rama con hojas pilíferas, ca. 2.5–4 mm de longitud   *S. leucotrichum*
1. Rama con hojas agudas o apiculadas, ca. 0.8–1.5 mm de longitud
          S. nigricans

Squamidium leucotrichum (Tayl.) Broth., Nat. Pfl. 1(3): 809. 1906. (Fig. 27A).

Napo: Near Tena, 6 km along Río Pano, 00°58'S, 77°52'W, 600 m, *Holm-Nielsen & Jeppesen 682* (AAU; Robinson *et al.*, 1971). Illustration based on *Holm-Nielsen & Jeppesen 1279* (AAU) from Ecuador.

In Ecuador reported also from Chimboraza, Cotopaxi, Pichincha and Tungurahua; at elevations from 600–3080 m.

Squamidium nigricans (Hook.) Broth., Nat. Pfl. 1(3): 808. 1906. (Fig. 27B).

Napo: Shinguipino, between Río Napo and Tena, ca. 460 m, *Grubb et al. 2933, 2936b* (Bartram, 1964). Illustration based on *Holm-Nielsen et al. 4530* (AAU) from Ecuador.

In Ecuador reported from Azuay, Chimboraza, Loja, Morona-Santiago, Pastaza, Pichincha and Tungurahua; at elevations from 460–3500 m.

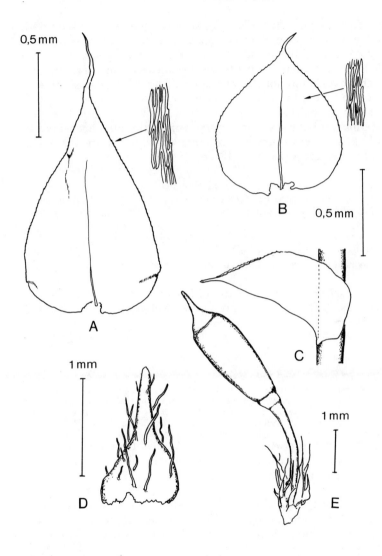

Figure 28. A. *Zelometeorium patulum*, A. leaf; B-E. *Zelometeorium recurvifolium*, B. leaf, C. side view of leaf sheathing stem, D. calyptra, E. sporophyte.

Zelometeorium Manuel

Secondary stems often pendent; leaves squarrose to recurved, broadly ovate to ovate-lanceolate, apex acute to acuminate, base cordate, clasping the stem, margins entire to mostly serrate, costae 1/2–3/4 lamina length, median cells linear-vermicular, smooth, branch leaves often not clasping at base; dioicous; setae short, seta and capsule length ±equal; capsules ovoid-cylindrical; calyptrae short mitrate, hairy.

Zelometeorium is commonly epiphytic but occasionally epiphyllus or less often lignicolous. Four species are known in Ecuador. Reference: Manuel (1977b).

Key to the Species

1.	Leaves ovate-lanceolate, long acuminate	*Z. patulum*
1.	Leaves broadly ovate, short acuminate	*Z. recurvifolium*

Clave para las Especies

1.	Hojas ovado-lanceoladas, largamente acuminadas	*Z. patulum*
1.	Hojas ampliamente ovadas, cortamente acuminadas	*Z. recurvifolium*

Zelometeorium patulum (Hedw.) Manuel, J. Hattori Bot. Lab. 43: 118. 1977. (Fig. 28A). Synonym: *Meteoriopsis patula* (Hedw.) Broth.

Napo: Añangu, ca. 75 km east of Coca, ca. 00°32'S, 76°23'W, 245–325 m, *Churchill & Sastre-De Jesús 13802* (NY; Churchill *et al.*, 1992); At Río Suno, 3 km W of Río Napo, 00°42'S, 77°10'W, 400 m, *Holm-Nielsen & Jeppesen 931* (AAU); Mishuallí, near junction of Río Mishuallí & Río Napo, 01°03'S, 77°41'W, 500 m, *Holm-Nielsen 19268* (AAU). **Pastaza:** Río Bobonaza, 460 m, *Spruce 1244* (NY; Mitten, 1869); Yuralpa, 00°55'S, 77°21'W, 440 m, *Holm-Nielsen & Jeppesen 942* (AAU; Robinson *et al.*, 1971); Curaray, northern bank 2 km W of school, 01°22'S, 76°58'W, 250 m, *Holm-Nielsen et al. 21871* (AAU); Montalvo, near military camp, 02°05'S, 76°58'W, ca. 250 m, *Løjtnant & Molau 13512* (AAU). **Sucumbíos:** Reserva Faunística Cuyabeno, near Laguna Grande, 00°00', 76°12'W, 265 m, *Balslev 84918* (AAU); *Heikkinen RH-1990-1 & -27* (NY). Illustration based on *Balslev 84918*.

In Ecuador reported also from Azuay, Chimborazo, Loja, Morona-Santiago, Los Ríos, and Pichincha, also the Galapagos Islands; at elevations from 245–1220 m.

Zelometeorium recurvifolium (Hornsch.) Manuel, J. Hattori Bot. Lab. 43: 121. 1977. (Fig. 28B–E). Synonyms: *Meteorium onustum* Spruce ex Mitt., *Meteoriopsis recurvifolia* (Hornsch.) Broth.

Napo: El Napo, 500 m, *Benoist 4664* (Thériot, 1936); Shinguipino, between Río Napo and Tena, ca. 460 m, *Grubb et al. 2932* (Bartram, 1964); Río Payamino, 00°29'S, 77°12'W, 350 m, *Holm-Nielsen & Jeppesen 762* (AAU; Robinson et al., 1971); *Holm-Nielsen & Jeppesen 838* (AAU); Añangu, ca. 75 km east of Coca, ca. 00°32'S, 76°23'W, 245–325 m, *Churchill & Sastre-De Jesús 13791-a* [c.fr.], *13795, 13807, 13864, 13879* (AAU, NY; Churchill *et al.*, 1992); *Lawesson et al. 39734* (AAU); Tangoy, 1 hr downstream along Río Aguarico from Zancudo flooded forest,

00°34'S, 75°27'W, 290 m, *Holm-Nielsen et al. 20113* (AAU). **Pastaza:** Río Pastaza between Destacamento Chiriboga and Apachi Entza, ca. 02°20–32'S,76°55–77°08'W, 285 m, *Øllgaard et al. 35238* (AAU). Illustration based on *Churchill & Sastre-De Jesús 13791-a.*

In Ecuador reported also from Morona-Santiago; at elevations from 245–1000 m.

NECKERACEAE

Neckeropsis Reichdt.

Plants with primary stems creeping, secondary stems ascending, usually perpendicular to substrate; leaves complanate, smooth to undulate, broadly ovate-ligulate, asymmetric, apex truncate, base decurrent on one side, costae single, 3/4–4/5 lamina length, median cells irregularly rhomboidal or hexagonal, smooth; perichaetial leaves linear, often as long or longer than sporophyte; seta short: capsule erect, short-cylindrical, peristome double, exostome teeth 16, papillose, endostome basal membrane low, segments 16, cilia absent; opercula long rostraste; calyptrae mitrate-campanulate, smooth to hairy.

In Ecuador the Neckeraceae is represented by three genera: *Neckera* (including *Neckeradelphus*) with four or five species of montane forests, *Neckeropsis* of lowland tropical to premontane forests with two species and *Isodrepanium lentulum* (Wils.) Britt. Although not recorded for Amazonas, *Isodrepanium* is likely to be found in the transition from premontane to lowland forests. It is characterized by the complanate habit with the leaves cultriform (falcate, distal half of lamina strongly arched back), margins serrulate, costae absent or weak, short and forked (Fig. 33b; illustration based on *Sastre-De Jesús et al. 1268* (NY) from Colombia). *Porotrichum* (including *Porothamnium*) and *Pinnatella* have traditionally been placed in the Neckeraceae, but are now placed in the Thamnobryaceae following Sastre-De Jesús (1987).

Both species of *Neckeropsis* are characteristic elements of the tropical lowlands, occurring as epiphytes on trunks and branches of trees and shrubs, and on lianas.

Key to the Species

1. Leaves smooth, never auriculate   *N. disticha*
1. Leaves undulate, particularly notable when dry, often auriculate
      N. undulata

Clave para las Especies

1. Hojas lisas, nunca auriculadas   *N. disticha*

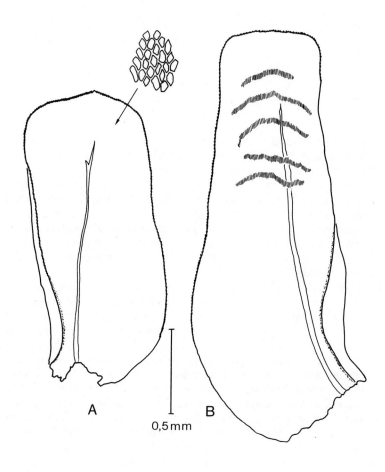

Figure 29. A. *Neckeropsis disticha*, A. leaf; B. *Neckeropsis undulata*, B. leaf.

1.   Hojas onduladas, particularmente notable cuando están secas, a menudo auriculadas   ***N. undulata***

Neckeropsis disticha (Hedw.) Kindb., Canad. Rec. Sc. 6: 21. 1894. (Fig. 29A).

Napo: Shinguipino, between Río Napo and Tena, ca. 460 m, *Grubb et al. 2921a* (Bartram, 1964); Río Payamino, ca. 10 km from mouth, 250 m, *Fransén 63* (AAU, GB); Añangu, ca. 75 km east of Coca, 00°32'S, 76°23'W, 245–325 m, *Churchill & Sastre-De Jesús 13789, 13791-e, 13838, 13868* (AAU, NY; Churchill *et al.*, 1992); ca. 24 km S of Tena on road to Puyo, ca. 01°12'S, 77°51'W, 600 m, *Churchill & Sastre-De Jesús 13670* (NY). **Sucumbíos:** Reserva Faunística Cuyabeno, N of Laguna Grande, 00°01'N, 76°11'W, 265 m, *Heikkinen RH-1990-88* (NY). Illustration based on *Fransén 63*.

In Ecuador reported also from Loja, Los Ríos, and Morona-Santiago; at elevations from 245–1000 m.

Neckeropsis undulata (Hedw.) Reichdt., Reise Oesterr. Freg. Novara Bot. 1: 181. 1870. (Fig. 29B).

Morona-Santiago: Yurupaz, 600 m, *Harling 2245b* (Crum, 1957). **Napo:** Shinguipino, between Río Napo and Tena, ca. 460 m, *Grubb et al. 2921* (Bartram, 1964); Coca–Armenia Vieja road, 15 km south of Coca, 250 m, *Fransén 48* (GB); Añangu, ca. 75 km east of Coca, ca. 00°32'S, 76°23'W, 245–325 m, *Churchill & Sastre-De Jesús 13783, 13785, 13793, 13866* (AAU, NY; Churchill *et al.*, 1992). **Pastaza:** Montalvo, 2°05'S, 76°58'W, ca. 250 m, *Løjtnant & Molau 13362* (AAU). Illustration based on *Churchill & Sastre-De Jesús 13793*.

In Ecuador reported also from Chimborazo, Loja, Los Ríos, and Tungurahua; at elevations from 245–1000 m.

ORTHOTRICHACEAE

Plants forming creeping mats; primary stems creeping, usually densely tomentose; secondary stems short, erect, occasionally tomentose, leaves often contorted or crispate when dry, spreading when wet, oblong-lingulate or lanceolate, costae single, strong, median cells isodiametric, smooth or mamillose, basal cells often elongate and papillose or mamillose; setae erect, elongate, smooth, capsules erect, urn ovoid to ovoid-oblong, peristome double, or appearing single, exostome teeth 16; opercula rostrate; calyptrae mitrate-campanulate, smooth or plicate, naked or hairy, deeply divided at base.

A large family in the Neotropics, and in Ecuador with possibly 70 species in seven genera, particularly of the montane regions. Our few species in Amazonas are found as epiphytes on trunks and branches of trees and shrubs (probably more likely to be found in the canopy), and occasionally on logs. Reference: Grout (1946).

Key to the Genera

1. Leaf base bordered by several rows of linear cells, inner cells oval-rounded; calyptrae smooth   *Groutiella*
1. Leaf base not bordered, marginal and inner cells elongate; calyptrae smooth or plicate   **2.**
    2. Lower and basal cells not papillose (tuberculate); calyptrae campanulate, smooth   *Schlotheima*
    2. Lower and basal cells papillose (tuberculate); calyptrae mitrate, plicate   *Macromitrium*

Clave para los Géneros

1. Base de la hoja bordeada por varias filas de células lineares, células al interior ovalado-redondeadas; caliptra lisa   *Groutiella*
1. Base de la hoja no bordeada, células marginales y al interior elongadas; caliptra lisa o plicada   **2.**
    2. Células inferiores y basales no papilosas (tuberculadas); caliptra campanulada, lisa   *Schlotheima*
    2. Células inferiores y basales papilosas (tuberculadas); caliptra mitriforme, plicada   *Macromitrium*

Groutiella Steere

Groutiella tomentosa (Hornsch.) Wijk & Marg., Taxon 9: 51. 1960. (Fig. 30A).

Secondary stem leaves oblong-lanceolate, 1.8–2.2 mm long, 0.3–0.4 mm wide, margins entire, costae percurrent, median cells isodiametric, mamillose, lower and basal cells oval-round, mamillose, marginal cells elongate and forming a border at base; peristome reduced or appearing absent - sporophytes not observed.

Six species of *Groutiella* are presently recorded for Ecuador.

Napo: Añangu, ca. 75 km east of Coca, ca. 00°32'S, 76°23'W, 245–325 m, *Churchill & Sastre-De Jesús 13860* (AAU, NY; Churchill *et al.*, 1992). **Sucumbíos:** Reserva Faunística Cuyabeno, N of Laguna Grande, 00°01'N, 76°11'W, 265 m, *Heikkinen RH-1990-332c* (NY). Illustration based on *Churchill & Sastre-De Jesús 13860.*

Macromitrium Brid.

Primary stems creeping, secondary stems short to ±elongate, ascending to erect, leaves often crispate when dry, lanceolate to oblong-ligulate, short acuminate or obtuse-rounded, costae percurrent, median cells mostly oval or round, thick-walled, weakly to somewhat strongly mamillose, lower and basal cells elongate, rectangular, tuberculate, papillae single, over cell lumen, particularly noticable along folds or plications; calyptrae plicate.

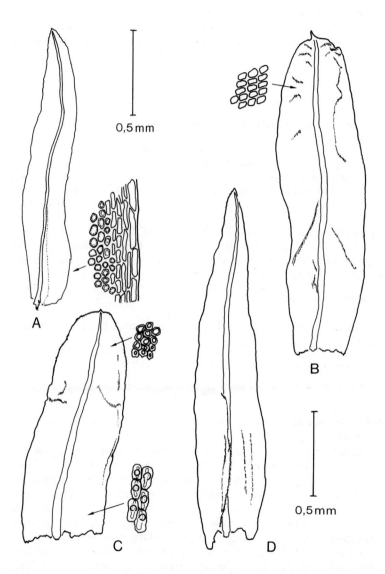

Figure 30. A. *Groutiella tomentosa,* A. leaf; B. *Schlotheimia torquata,* B. leaf; C. *Macromitrium stellulatum,* C. leaf; D. *Macromitrium portoricense,* D. leaf.

Just over 30 species of *Macromitrium* are listed for Ecuador. The genus is one of the largest genera for the country, and typically found as epiphytes on trunks and branches of trees in premontane to montane forests.

Key to the Species

1. Leaves ±narrowly lanceolate, ca. 2 mm long, apex short acuminate
   M. portoricense
1. Leaves oblong-ligulate ca. 3 mm long, apex obtuse   *M. stellulatum*

Clave para las Especies

1. Hojas ±estrechamente lanceoladas, ca. 2 mm de longitud, ápice cortamente acuminado   *M. portoricense*
1. Hojas oblongo-liguladas, ca. 3 mm de longitud, ápice obtuso
   M. stellulatum

M. portoricense Williams, Bryologist 32: 69. 1929. (Fig. 30D).
Our plants match rather well the description given by Florschütz (1964), however there very likely exists an earlier name than Williams taxon, thus the name is only tentitively used.
Pastaza: Montalvo, N and E of military camp, 02°05'S, 76°58'W, ca. 250 m, *Løjtnant & Molau 13384* (AAU). Illustration based on *Løjtnant & Molau 13384*.

M. stellulatum (Hornsch.) Brid., Bryol. Univ. 1: 314. 1826. (Fig. 30C).
Pastaza: Montalvo, N and E of military camp, 02°05'S, 76°58'W, ca. 250 m, *Løjtnant & Molau 13344*(AAU). Illustration based on *Løjtnant & Molau 13344*.

Schlotheimia Brid.

Schlotheimia torquata (Hedw.) Brid., Spec. Musc. 2: 16. 1812. (Fig. 30B).
Secondary stem 1.5–2.0 cm tall, leaves ±spirally appressed, crispate, ovate- to oblong-lingulate, 2.5–3.0 mm long, 0.5–0.7 mm wide, rugose, obtuse-rounded and apiculate, margins plane to recurved below, entire, costae precurrent, median cells oval to subquadrate, thick-walled, smooth, basal and marginal cells similar, elongate, cell angles appearing papillose, porose; perichaetial leaves 2 × longer than secondary stem leaves; setae 6–7 mm long, urn 1.8–2.0 mm long, plicate, exostome papillose, endostome ±reduced; calyptrae campanulate, ca. 3.5 mm long, smooth below, distally scabrous.
Schlotheimia contains three or four species in Ecuador.
Napo: Near Tena, 6 km along Río Pano, 00°58'S, 77°52'W, 600 m, *Holm-Nielsen & Jeppesen 694* (AAU; Robinson et al., 1971). Illustration based on *Holm-Nielsen & Jeppesen 694*.

PHYLLODREPANIACEAE

The Phyllodrepaniaceae contain two genera with a wide neotropical distribution. Although not recorded for Ecuador, *Phyllodrepanium falcifolium* (Schwaegr.) Crosby (*Drepanophyllum*) is likely to be found in Amazonas; it is distinguished from *Mniomalia* in the following key, see also Figure 31 (illustration based on *Spruce 629* [BM; with *Callicostella rufescens*] from Brazil).

Key to the Genera

1. Plants dark and dull green; median cells isodiametric, papillose
   Mniomalia
1. Plants glossy green; median cells rhomboidal to rhombic, smooth
   *Phyllodrepanium**

Clave para los Géneros

1. Plantas verde-opacas y oscuras; células medias isodiamétricas, papilosas
   Mniomalia
1. Plantas verde brillantes; células medias romboidales a rómbicas, lisas
   *Phyllodrepanium**

Mniomalia C. Müll.

Mniomalia viridis (Mitt.) C. Müll., J. Mus. Godeffroy 3(6): 61. 1874. (Fig. 34A).

Plants small, solitary or forming short turfs; stems solitary or few branched, to 2 cm tall, leaves appearing 4-ranked, complanate, asymmetric, oblong, 0.6–1.2 mm long, ca. 0.3 mm wide, acute to rounded-apiculate, margins recurved, serrulate in distal 1/3 of lamina, costae single, percurrent, strong, median cells isodiametric, papillose; cylindrical propagula clustered on terminal stalk of stems or branches; sporophytes not observed.

A single widespread species in the neotropical lowlands, typically on tree trunks.

Pastaza: Río Bobonaza, 365 m, *Spruce 557* (NY; Mitten, 1869). Illustration based on *Churchill et al. 16125* (NY) from Colombia.

PHYLLOGONIACEAE

Phyllogonium Brid.

Phyllogonium fulgens (Hedw.) Brid., Bryol. Univ. 2: 671. 1827. (Fig. 31B).

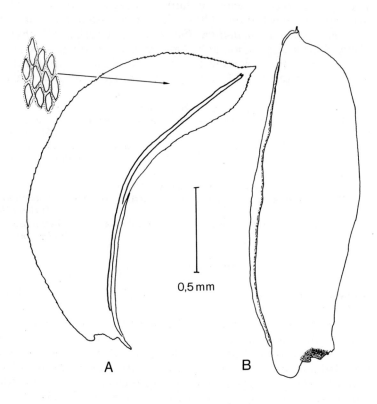

Figure 31. A. *Phyllodrepanium   falcifolium*, A. leaf; B. *Phyllogonium fulgens*, B. side view of leaf.

Plants with primary stems creeping, secondary stems often pendent; leaves appearing 2-ranked, oblong, concave and folded flat (U-shaped in cross-section), 2–3 mm long, 1 mm wide, apex short apiculate, base auriculate, margins smooth, ecostate or indistinctly short and forked, median cells linear, smooth, porose, alar cells short rectangular to oval, porose; setae short, ca. 2–3 mm long; capsules ovoid, ca. 1.2 mm long; opercula short rostrate; calyptrae cucullate, sparsely hairy.

Phyllogonium is typically epiphytic and pendent in habit, more common in the premontane and montane forests in Ecuador. The only other species in Ecuador is *P. viscosum* (P. Beauv.) Mitt. usually more common at higher elevations with longer leaves (4–5 mm) that are undulate, and apices truncate. Reference: Lin (1983).

Napo: Near Tena, 6 km along Río Pano, 77°52'W, 00°58'S, 600 m, *Holm-Nielsen & Jeppesen 700* (AAU; Robinson *et al.*, 1971). Illustration based on *Holm-Nielsen & Jeppesen 557* (AAU) from Ecuador.

In Ecuador reported also from Loja, Morona-Santiago, and Pichincha; at elevations from 600–2000 m.

POTTIACEAE

Plants erect, forming turfs, occasionally gregarious; leaves linear-lanceolate to ovate or obovate but mostly broadly to narrowly lanceolate, ovate-lanceolate, margins plane, recurved or incurved, entire or serrate distally, border present or absent, costae single, percurrent to long excurrent, in cross-section costa with 1 or 2 stereid bands, median cells small, isodiametric, mostly thick-walled, smooth, mamillose or papillose; sporophyte terminal, setae mostly elongate, smooth, often twisted, capsule mostly erect, urn cylindrical or ovoid-cylindrical, rarely curved, peristome single, fragile, teeth 16, often divided into 32 filaments, usually papillose; opercula mostly rostrate; calyptrae cucullate, naked and smooth; spores spherical, usually papillose.

A very large family in Ecuador with some 33 genera and over 100 species recorded. While some species are found in montane forests, most are found in disturbed open montane areas, or associated with xerophytic grassland and shrub vegetation, and páramo .

Key to the Genera

1. Leaf margins distinctly bordered; median cells mostly hexagonal, thin-walled   *Dolotortula*

1.   Leaf margins not bordered; median cells isodiametric, thick-walled   **2.**
     2.   Leaf margins entire; leaves ovate-lanceolate to oblong-lingulate
                                                                     Barbula
     2.   Leaf margins distally ±serrate, often irregularly so; leaves oblong-
          obovate to spathulate   *Hyophila*

Clave para los Géneros
1.   Margen de la hoja bordeada; células medias en su mayoría hexagonales
     pared delgada   *Dolotortula*
1.   Margen de la hoja no bordeada; céllulas medias isodiametricas, pared
     gruesa   **2.**
     2.   Margenes de la hoja enteros; hojas ovado-lanceoladas hasta
          oblongo-liguladas   *Barbula*
     2.   Margenes de la hoja distalmente ±serrados, a menudo
          irregularmente; hojas oblongo-obovadas hasta espatuladas
                                                                   Hyophila

Barbula Hedw.

Plants small; leaves contorted when dry, erect-spreading when
wet, lanceolate or triangular-subulate to ligulate, margins
recurved below or throughout, entire, costae percurrent to short
excurrent, in cross-section stereids above and below guide cells,
median cells oval to subquadrate, thick-walled, smooth to
papillose, basal cells short to ±long rectangular; dioicous; setae
elongate; capsules erect, urn ovoid-cylindrical, peristome usually
rather deeply divided, papillose and spirally twisted; opercula
rostrate.

Barbula is generally found on exposed soil. The genus contains
about 16 species recorded for Ecuador, however *Barbula* is very
poorly understood in the Andes and is in need of revision. A number
of species previously placed in *Barbula* have recently been
transfered to *Didymodon* and *Pseudocrossidium* by Zander (1979,
1981a).

Key to the Species
1.   Leaves lanceolate-subulate, margins recurved throughout, median
     cells smooth, propagula absent   *B. arcuata*
1.   Leaves oblong-lingulate, margins plan distally, recurved below,
     median cells papillose, propagula often present in distal leaf axils
                                                                 B. indica

Clave para las Especies
1.   Hojas lanceolado-subuladas, márgenes recurvados en toda su longitud,
     células medias lisas, propágulos ausentes   *B. arcuata*

1. Hojas oblongo-liguladas, márgenes distalmente planos, recurvados hacia la parte inferior, células medias papilosas, propágulos a menudo presentes en los ejes distales de la hoja **B. indica**

Barbula arcuata Griff., Calcutta J. Nat. Hist. 2: 491. 1842. (Fig. 32A). Synonym: *Barbula subulifolia* Sull.

Pastaza: Río Bobonaza, 365 m, *Spruce 181* (NY; Mitten, 1869); near Mera, ca. 450 m, *Mexía 6960a* (NY). Illustration based on *Schultes s.n.* (NY) from Colombia.

In Ecuador reported also from Tungurahua; at elevations from 365–2155 m.

Barbula indica (Hook.) Spreng. in Steud., Nomencl. Bot. 2: 72. 1842. (Fig. 32B-C). Synonym: *Barbula cruegeri* Sond. ex C. Müll.

Morona-Santiago: Patuca, 600 m, *Harling 2291* (S; Crum, 1957). **Napo:** Añangu, ca. 75 km east of Coca, ca. 00°32'S, 76°23'W, 245–325 m, *Churchill & Sastre-De Jesús 13834, 13851* (AAU, NY; Churchill *et al.*, 1992); Mishuallí, near junction of Río Mishuallí & Río Napo, 01°03'S, 77°41'W, 500 m, *Holm-Nielsen 19284* (AAU). Illustration based on *Churchill & Sastre-De Jesús 13834*.

In Ecuador reported also from Tungurahua; at elevations from 245–1845 m.

Dolotortula Zander

Dolotortula mniifolia (Sull.) Zander, Phytologia 65: 426. 1989. (Fig. 5C). Synonym: *Tortula mniifolia* (Sull.) Mitt.

Plants medium sized, dull dark green to brownish-green; stems erect, to 1 cm tall, weakly radiculose; leaves loosely erect and contorted when dry, erect- to wide-spreading when wet, ligulate to more commonly spathulate, 4–5.5 mm long, to 1.5 mm wide, obtuse-rounded, bluntly apiculate, margins plane or weakly recurved below midleaf, entire and bordered, costae single, subpercurrent to percurrent, upper and median lamina cells oblong-hexagonal to rectangular-rounded, 30–40 µm long, 20 µm wide, thin-walled, smooth, marginal cells long linear, basal cells rectangular; dioicous; setae elongate, smooth; capsules erect, urn cylindrical, peristome single, divided into 32 filaments, papillose; opercula bluntly conic; spores lightly papillose.

Napo: Añangu, ca. 75 km east of Coca, ca. 00°32'S, 76°23'W, 245–325 m, *Churchill & Sastre-De Jesús 13845* (AAU, NY; Churchill *et al.*, 1992). Illustration based on *Churchill & Sastre-De Jesús 13845*.

Hyophila Brid.

Hyophila involuta (Hook.) Jaeg., Ber. St. Gall. Naturw. Ges. 1871–
72: 354. 1873. (Fig. 32D-E). Synonym: *Hyophila tortula*
(Schwaegr.) Hampe.

Plants small to medium sized, dull green, often blackish- or
brownish-green; leaves oblong-obovate to ±spatulate, 2–2.5 mm
long, apex acute to obtuse, often apiculate or mucronate, margins
entire below, distally irregularly serrate to dentate, occasionally
entire, costae percurrent, median cells irregularly quadrate to oval,
4–6 µm in diameter, mammillose, basal cells short rectangular,
±thin-walled and lax; propagula occasionaly present in leaf axil,
cylindrical propagula produced terminally on stalked cells; setae
elongate, smooth; capsule erect, peristome absent, opercula conic-
rostrate.

Hyophila involuta is more common at higher elevations often
found on soil covered rocks and concrete. In Ecuador three species
are recorded for *Hyophila*.

Napo: Shinguipino, between Río Napo and Tena, ca. 460 m, *Grubb et al. 2952* (Bartram,
1964). Illustration based on *Holm-Nielsen et al. 3132* (AAU) from Ecuador.

In Ecuador reported from Azuay, Tungurahua, and also the
Galapagos Islands; at elevations from 460–2700 m.

PTEROBRYACEAE

Plants with primary stems creeping, often wiry; secondary stems
pendent, frondose, or simple, leaves typically radially foliate or
spiral in 5 rows, concave, base auriculate, costae single (ecostate in
Hildebrandtiella), median cells linear to oblong-linear, smooth to
variously papillose, porose or not, alar region differentiated, cells
quadrate to subquadrate, often porose; dioicous; sporophytes
lateral, setae short to elongate, smooth, capsules erect, peristome
double, exostome smooth, endostome reduced; opercula short to
long rostrate; calyptrae cucullate and hairy or mitrate and smooth.

Pterobryaceae are a frequent element in the tropical lowland
and premontane forests, commonly found as epiphytes, on trees,
treelets and lianas. Although not recorded for Ecuador, *Jaegerina
scariosa* (Lor.) Arz. may be found in Amazonas and somewhat
higher elevations in premontane forests (Fig. 33a; illustration based
on *Churchill et al. 14781* (NY) from Colombia). Reference: Arzeni
(1954).

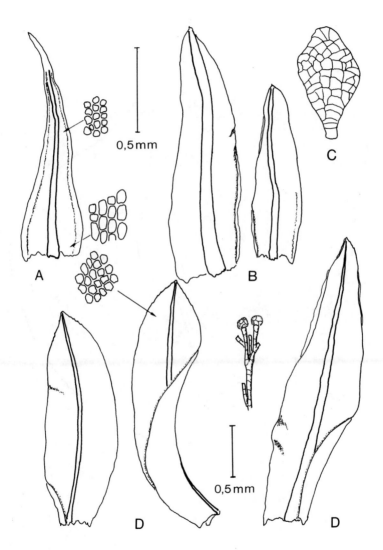

Figure 32. A. *Barbula arcuata*, A. leaf; B-C. *Barbula indica*, C. leaves, C. propagula; D-E. *Hyophila involuta*, D. leaves, E. stalked propagulae.

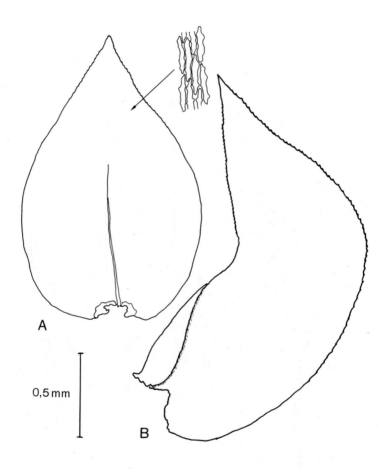

Figure 33. A. *Jaegerina scariosa*, A. leaf; B. *Isodrepanium lentulum*, B. leaf.

Key to the Genera

1. Secondary stems pendulous; leaves in 5-spiraled rows   **2.**
   2. Leaves ecostate; apex acuminate   *Hildebrandtiella*
   2. Leaves costate; apex long acuminate to piliferous   *Orthostichopsis*
1. Secondary stems erect, tufted or frondose; leaves not distinctly spiraled
                                                                3.
   3. Secondary stems frondose; leaves erect-spreading   *Pireella*
   3. Secondary stems simple, occasionally few branched   **4.**
      4. Leaves erect-spreading to wide-spreading; lamina cells
         with 1–2 papillae over cell lumen   *Henicodium*
      4. Leaves squarrose-spreading; lamina cells appearing
         smooth or projecting at distal angles   *Jaegerina**

Clave para los Géneros

1. Tallos secundarios péndulos; hojas en cinco filas espiraladas   **2.**
   2. Hojas sin costa; ápice acuminado   *Hildebrandtiella*
   2. Hojas con costa; ápice largamente acuminado hasta pilífero
                                                       Orthostichopsis
1. Tallos secundarios erectos, arracimados o frondosos; hojas no
   distintivamente espiraladas   **3.**
   3. Tallos secundarios frondosos; hojas erectas bien separadas
      del tallo   *Pireella*
   3. Tallos secundarios simples; ocasionalmente poco
      ramificados   **4.**
      4. Hojas erectas-separadas hasta ampliamente separadas
         del tallo; células de la lámina con 1–2 papilas sobre el
         lumen celular   *Henicodium*
      4. Hojas escuarrosas separadas del tallo; células de la
         lámina lisas o proyectandose en los ángulos distales
                                                       *Jaegerina**

Henicodium (C. Müll.) Kindb.

Henicodium geniculatum (Mitt.) Buck, Bryologist 92: 534. 1989.
(Fig. 34B).

Plants glossy green to yellowish-green, small, forming loose short tufts; primary stems creeping, tomentose, secondary stems mostly simple, rarely branched, distally often attenuated, leaves erect or appressed when dry, erect-spreading to spreading when wet, ovate- to oblong-lanceolate, 1.4–1.7 mm long, somewhat concave, plicate when dry, clasping at base, rather bluntly acuminate to acute, slightly decurrent, margins recurved, distally denticulate to serrulate, costae 2/3 lamina length, median cells linear-vermicular, papillose, papillae 1–2, appearing over cell lumen, often indistinct, insertion cells rectangular and porose, alar

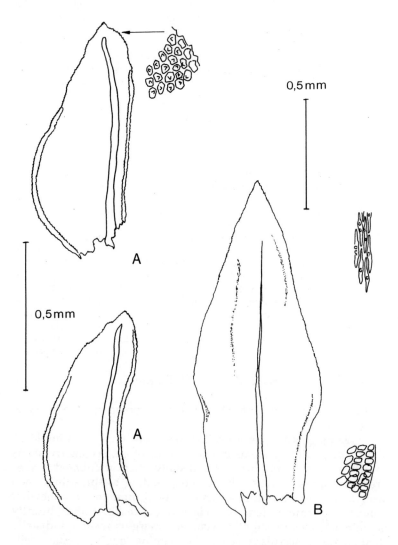

Figure 34. A. *Mniomallia viridis*, A. leaves; B. *Henicodon geniculatum*, B. leaf.

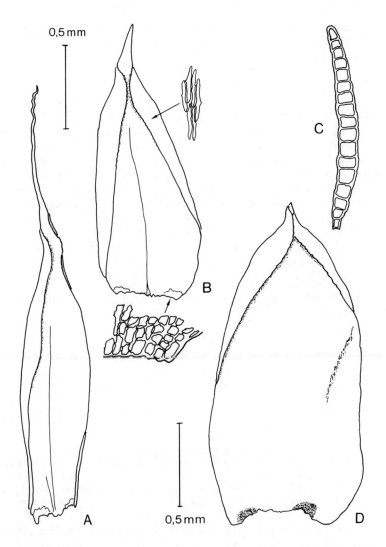

Figure 35. A-B. *Orthostichopsis praetermissa,* A. branch leaf, B. stem leaf; C-D. *Hildebrandtiella guyanensis,* C. propagula, D. leaf.

cells oblate rectangular or quadrate; dioicous; sporophytes not observed.

Sucumbíos: Reserva Faunística Cuyabeno, N of Laguna Grande, 00°01'N, 76°11'W, 265 m, *Heikkinen RH-1990-332a* (NY). Illustration based on *Heikkinen RH-1990-332a*.

Epiphytic on tree trunks, from moist to semi-dry lowland to premontane forests. *Leucodontopsis* is now replaced by an older name - *Henicodium*. Reference: Buck, W. R. 1989.

Hildebrandtiella C. Müll.

Hildebrandtiella guyanensis (Mont.) Buck, Brittonia 43: 97. 1991 (Fig. 35C-D). Synonyms: *Orthostichidium*, *O. excavatum* (Mitt.) Broth., *O. pentagonum* (Hampe & Lor.) C. Müll.

Plants glossy green to golden-green, forming mats; secondary stems ±robust, pendent, leaves in 5 spiral rows, 1.5–3 mm long, ca. 1–1.2 mm wide, concave, apex acuminate, margins entire to indistinctly serrulate distally, ecostate, median cells oblong-linear, smooth, porose, alar region not well developed, cells quadrate-rounded, porose, insertion cells golden-brown; propagula present in leaf axials, cylindrical; setae short, urn short cylindrical, calyptrae mitrate, smooth.

Napo: Shinguipino, between Río Napo and Tena, ca. 460 m, *Grubb et al. 2934* (Bartram, 1964). Illustration based on *Holm-Nielsen et al. 4529* (AAU) from Ecuador.

In Ecuador reported also from Loja and Morona-Santiago; at elevations from 460–1000 m.

Orthostichopsis Broth.

Key to the Species

1. Secondary stem leaves smooth; alar cells few, ca. 4–10 along margin
O. praetermissa
1. Secondary stem leaves distinctly plicate; alar cells numerous, ca. 20–30 along margin
*O. tetragona**

Clave para las Especies

1. Hojas del tallo secundario lisas, pocas células alares, ca. 2–10 a lo largo del márgen
O. praetermissa
1. Hojas del tallo secundario distintivamente plicadas, numerosas células alares, ca. 20–30 a lo largo del margen
*O. tetragona**

Orthostichopsis praetermissa Buck, Brittonia 43: 98. 1991. (Fig. 35A-B).

Plants glossy green or golden-green, forming mats; secondary stems pendent, leaves in 5 spiral rows, oblong abruptly ±short to long acuminate, 1.8–2.2 mm long, to 1 mm wide, lamina smooth, margins incurved distally, serrulate in distal 1/3 of lamina, costae weak, 2/3–3/4 lamina length, median cells linear-vermicular, porose, alar region differentiated, ca. 4–10 rows of quadrate cells along margin, insertion cells often golden-brown; distal branches often attenuated and elongate, branch leaves narrowly oblong & abruptly acuminate with a short to long hair point (1/2 as long as leaf proper); setae short, capsules oval to ovoid-oval, opercula short rostrate.

The report by Robinson *et al.* (1971) of *Orthostichopsis auricosta* (C. Müll.) Broth. is a likely synonym of *O. praetermissa*. The name *Orthostichopsis crinita* (Sull.) Broth. has been applied to the lowland Northern Andean countries but actually represents a different element (Buck, 1991). Possibly only two species are to be found in Ecuador. *Orthostichopsis tetragona* (Hedw.) Broth. is known from the lowlands of western Ecuador (Prov. Esmeraldas, Río Onzolé at confluence with Bellavista, 130 m, *Holm-Nielsen et al. 25839*, NY). Another feature that can aid to distinguish *O. praetermissa* from *O. tetragona* are the alar cells which are few in the former, and numerous in the latter species.

Morona-Santiago: Taisha, ca. 5 km N-NW of military camp, 02°23'S, 77°30'W, 500 m, *Brandbyge & Asanza C. 31849* (AAU). **Napo:** Añangu, ca. 75 km east of Coca, ca. 00°32'S, 76°23'W, 245–325 m, *Churchill & Sastre-De Jesús 13847, 13848, 13856, 13867* (NY; Churchill *et al.*, 1992); Near Tena, 6 km along Río Pano, 00°58'S, 77°52'W, 600 m, *Holm-Nielsen & Jeppesen 693, 696, 700* (AAU; Robinson *et al.*, 1971); near Sarsayaco village, Río Negro, 500 m, *Balázs 69-16/M* (NY). Illustration based on *Churchill & Sastre-De Jesús 13856.*

In Ecuador reported also from Pastaza and Tungurahua; at elevations from 245–2155 m.

Pireella Card.

Pireella pohlii (Schwaegr.) Card., Rev. Bryol. 40: 18. 1913. (Fig. 36C-F).

Plants light to dark green, often forming loose or dense tufts; primary stem creeping, leaves scale-like; secondary stems erect, 3–4(5) cm tall, dendroid, with a distinct stipe, stipe leaves appressed, leaf apices slenderly long acuminate, stem and branch leaves loosely erect to erect-spreading, oblong-lanceolate, to 1.8 mm long, deeply concave, apex ±abruptly acuminate, ±flat, margins weakly serrulate usually throughout or entire, costae percurrent or ending below apex, median cells oblong-linear, appearing smooth or weakly papillose; dioicous; setae elongate, ca. 7–9 mm long, smooth or distally roughened, urn ovoid to ovoid-cylindrical, 1.4–1.8 mm long, opercula long rostrate; calyptrae sparsely hairy.

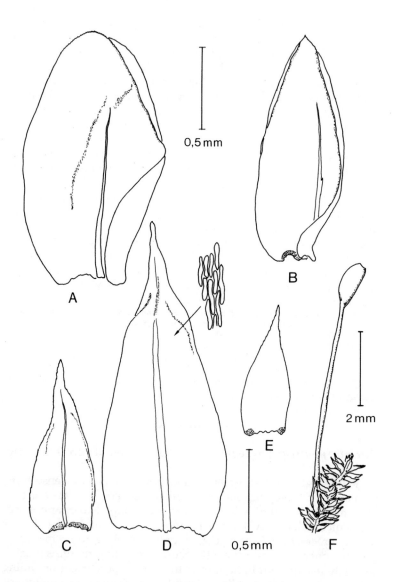

Figure 36. A-B. *Porotrichum lindigii*, A. stem leaf, B. branch leaf; C-F. *Pireella pohlii*, C. branch leaf, D. stem leaf, E. distal deciduous branch leaf, F. sporophyte.

Pireella in Ecuador is represented by four species of which one has been reported from Amazonas.

Morona-Santiago: Yurupaz, 600 m, *Harling 2259b, 2262b* (Crum, 1957). **Napo:** Shinguipino, between Río Napo and Tena, ca. 460 m, *Grubb et al. 2925c* (Bartram, 1964); SW of Nuevo Rocafuerte, along Río Braga, 200–230 m, *Jaramillo & Coello 4594* (AAU); Añangu, ca. 75 km east of Coca, ca. 00°32'S, 76°23'W, 245–325 m, *Churchill & Sastre-De Jesús 13810, 13861, 13864-b, 13878* (AAU, NY; Churchill *et al.*, 1992); near Tena, 6 km along Río Pano, 00° 58'S, 77°52'W, 400 m, *Holm-Nielsen & Jeppesen 926, 930* (AAU; Robinson *et al.*, 1971). Illustration based on *Churchill & Sastre-De Jesús 13864-b.*

Some forms encountered at Añangu appeared similar to *Pireella angustifolia* (Schwaegr.) Card. and thus labeled so based on the very narrow leaves. However the stem leaves are squarrose in *P. angustifolia,* a character not exhibited in the Añangu collections. Many of our collections exhibited deciduous distal branch leaves.

RACOPILACEAE

Racopilum P. Beauv.

Racopilum tomentosum (Hedw.) Brid., Bryol. Univ. 2: 720. 1827. (Fig. 37D-E).

Plants forming mats; primary stems creeping, often densely tomentose, secondary stems spreading, occasionally pendent, leaves dimorphic, upper (dorsal) leaves small, narrowly triangular, 1.5–2.0 mm long, narrowly acuminate, costae long excurrent, lateral leaves ovate-lancaolate to more commonly oblong, 2–3 mm long, 0.8 mm wide, acuminate to acute, margins plane, distally serrate, median cells irregularly isodiametric or rhomboidal, thick-walled, smooth to mammillate; setae 2.5 cm long, capsules often curved, urn cylindrical, striate; peristome double, exostome striate below, papillose above, endostome basal membrane high, 2–3 cilia; opercula rostrate; calyptrae cucullate, sparsely hairy.

More common at higher elevations, this species in Ecuador is commonly found in shaded or exposed places, on logs, branches of treelets or shrubs, occasionally on soil.

Morona-Santiago: Patuca, 600 m, *Harling 2273 , 2274* (S; Crum, 1957). **Napo:** Las Sachas, ca. 40 km NE of Coca, ca. 250 m, *Fransén 62* (AAU, GB); Añangu, ca. 75 km east of Coca, ca. 00°32'S, 76°23'W, 245–325 m, *Churchill & Sastre-De Jesús 13796* (NY; Churchill *et al.*, 1992). Illustration based on *Fransén 62.*

In Ecuador reported from Azuay, Chimborazo, Loja, Pastaza, Pichincha, and also the Galapagos Islands; at elevations from 245–3690 m.

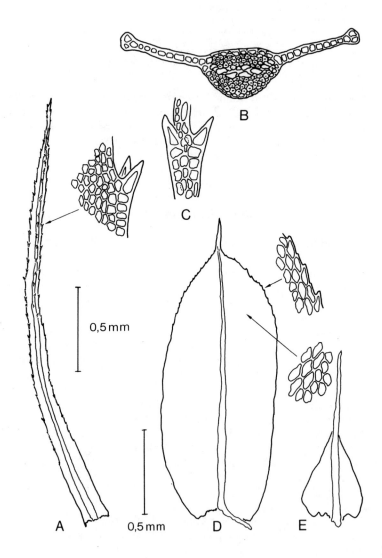

Figure 37. A-C. *Pyrrhobryum spiniforme*, A. leaf, B. x-section of leaf, C. side view of double toothed margin; D-E. *Racopilum tomentosum*, D. lateral leaf, E. dorsal (upper) leaf.

RHIZOGONIACEAE

Pyrrhobryum Mitt.

Pyrrhobryum spiniforme (Hedw.) Mitt., J. Linn. Soc., Bot. 10: 174. 1868. (Fig. 37A-C). Synonym: *Rhizogonium spiniforme* (Hedw.) Bruch.

Plants forming loose to dense tall tufts, to 6 cm tall, stems erect, wiry, tomentose at base; leaves linear to linear-lanceolate, 6–8 mm long, 0.5–0.7 mm wide, apex acuminate, margins bistratose, doubly-toothed, costae single, toothed on back, excurrent, median cells isodiametric, 4–6 sided; synoicous, inflorescence at base of stems; setae elongate; capsules inclined, urn cylindrical, curved, ±striate, peristome double, exostome striate below; opercula conic-rostrate, oblique; calyptrae cucullate, naked.

More common at slightly higher elevations in Ecuador, this species is commonly found on soil, decaying logs and occasionally on tree trunk bases. The only other species, *P. mnioides* (Hook.) Manuel, is found in high montane forests, and exhibits a distinct decurrent leaf base.

Morona-Santiago: Yurupaz, 600 m, *Harling 2266* (S; Crum, 1957). **Napo:** Trail from Gavilanes to Las Palmeras, 600 m, *Mexía 7067a* (GB). Illustration based on *Mexía 7067a*.

In Ecuador reported also from Loja; at elevations from 600–1000 m or more.

SEMATOPHYLLACEAE

Plants forming mostly mats or creeping tall turfs; stems spreading or ascending, leaves ovate to oblong or lanceolate, costae absent, rarely short and forked, median cells mostly oblong-linear to rhomboidal, smooth or papillose, alar cells differentiated, mostly oval to occasionally short rectangular, often inflated, golden-yellow or brown; sporophyte lateral, setae elongate, rarely short, smooth or distally roughened or papillose, capsule erect or inclined to pendulous, urn short ovoid, peristome double, exostome teeth 16, striate, furrowed or not, endostome basal membrane high, segments 16, cilia 1–2; opercula commonly long rostrate, oblique, or conic; calyptrae cucullate, smooth.

Sematophyllaceae are commonly found in shaded places on trunks of trees and treelets, and over logs.

Key to the Genera

1. Median leaf cells smooth **2.**

2. Leaf strongly concave distally, alar cells large, at ca. 45° angled
   Acroporium
2. Leaf slightly concave, alar cells not or slightly angled
   Sematophyllum
1. Median leaf cells papillose, papillae over cell lumen (see also
   Acroporium)　　　　　　　　　　　　　　　　　　　　**3.**
   3. Leaf cells pluripapillose, papillae several in a row　　*Taxithelium*
   3. Leaf cells with a single papillae over cell lumen　　*Trichosteleum*

Clave para los Géneros
1. Células medias de la hoja lisas　　　　　　　　　　　　**2.**
   2. Hoja fuertemente cóncava distalmente, células alares grandes,
      anguladas ca. 45°　　　　　　　　　　　　　*Acroporium*
   2. Hoja levemente cóncava, células alares levemente onduladas o
      no　　　　　　　　　　　　　　　　　　*Sematophyllum*
1. Células medias de la hoja papilosas, papilas sobre el lumen celular (ver
   Acroporium)　　　　　　　　　　　　　　　　　　　**3.**
   3. Células de la hoja pluripapilosas, varias papilas en una fila
      Taxithelium
   3. Células de la hoja con papilas simples sobre el lumen celular
      Trichosteleum

Acroporium Mitt.

Plants mostly glossy green to yellow or golden, forming ±tall tufts; stems ascending-erect; leaves ovate-lanceolate, 1.6–2.6 mm long, concave below, enrolled distally, apex short to long acuminate, base subauriculate, costae absent, median cells linear to linear-vermicular, smooth to unipapillose, porose or not, alar cells distinctly inflated, sharply angled (ca. 45°), golden-brown; autoicous; setae elongate, smooth, capsules inclined, urn ovoid, mouth flared, exostome furrowed; opercula long rostrate, oblique.

Key to the Species
1. Leaf lamina cells unipapillose, papillae over cell lumen　*A. guianense*
1. Leaf lamina cells smooth　　　　　　　　　　　　　*A. pungens*

Clave para las Especies
1. Células de la lámina unipapilosas, papilos sobre el lumen
   A. guianense
1. Células de la lámina lisas　　　　　　　　　　　　*A. pungens*

Acroporium guianense (Mitt.) Broth., Nat. Pfl. ed. 2, 11: 436. 1925.
　　Sucumbíos: Reserva Faunística Cuyabeno, N of Laguna Grande, 00°01'N, 76°11'W, 265 m, *Heikkinen RH-1990-43* (NY).

This species is considered a synonym of *Acroporium pungens* by Florschütz-de Waard (1990), however the collection cited above does not appear to be within the variation presently observed in Ecuadorean Amazonia. Until a more detailed study of this genus is completed for the Neotropics, this species is retained as distinct.

Acroporium pungens (Hedw.) Broth., Nat. Pfl. ed. 2, 11: 436. 1925. (Fig. 38A).

Napo: Near Tena, 6 km along Río Pano, 00°58'S, 77°52'W, 600 m, *Holm-Nielsen & Jeppesen 684* (AAU; Robinson *et al.*, 1971); Mishuallí, ca. junction of Río Mishuallí-Río Napo, 01°03'S, 77°41'W, 500 m, *Holm-Nielsen 19310* (AAU). **Pastaza:** Curaray, SE of airstrip, 01°22'S, 76°57'W, 250 m, *Holm-Nielsen et al. 22156* (AAU). **Sucumbíos:** Reserva Faunística Cuyabeno, N of Laguna Grande, 00°01'N, 76°11'W, 265 m, *Heikkinen RH-1990-290a* (NY). Illustration based on *Holm-Nielsen & Jeppesen 684*.

This species is commonly epiphytic, from wet lowland to montane elevations. In Ecuador reported also from Chimborazo, Los Ríos, Morona-Santiago and Pichincha; at elevations from 30–1100 m or more.

Sematophyllum Mitt.

Plants small to somewhat medium sized, forming mats; stems spreading to ascending and often homomallous, leaves ovate to ovate-lanceolate, concave, broadly or narrowly acuminate, margins plane to recurved, entire to serrulate distally, costae absent, median cells linear to oblong-fusiform, smooth, alar cells inflated, several, usually golden-brown; autoicous; setae elongate, smooth, capsules suberect to subpendulous, short ovoid, exostome striate; opercula rostrate, oblique.

A rather large and poorly known genus in the Neotropics, particularly in the Andean premontane and montane forests. Nine species are recorded for Ecuador.

Key to the Species

1. Plants forming tufts, irregularly branched, secondary stems and branches suberect; leaves broadly ovate or ovate-lanceolate, apex short acuminate   *S. subpinnatum*
1. Plants forming mats, subpinnate, secondary stems and branches spreading; leaves ovate-acuminate, apex acuminate   *S. subsimplex*

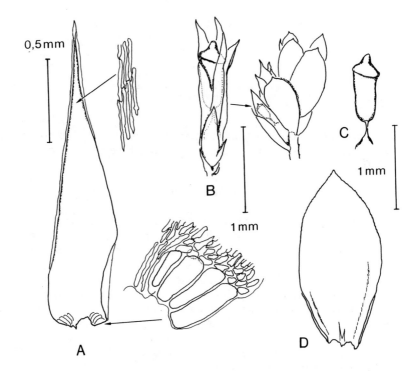

Figure 38. A. *Acroporium pungens*, A. leaf; B-D. *Hydropogon fontinaloides*, B. distal leafy branch with immersed sporophyte, C. capsule, D. leaf.

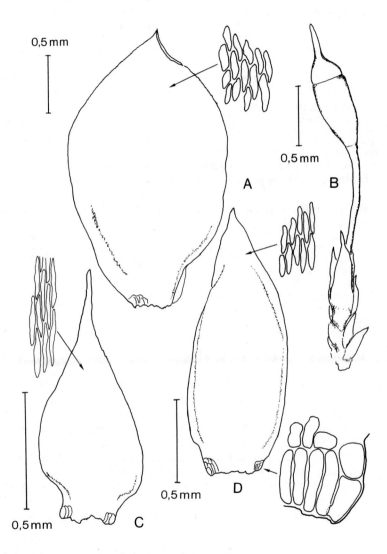

Figure 39. A-B. *Sematophyllum subpinnatum*, A. leaf, B. sporophyte; C. *Sematophyllum subsimplex*, C. leaf; D. *Semtophyllum subpinnatum*, D. leaf.

Clave para las Especies

1. Plantas formando racimos, irregularmente ramificadas, tallos secundarios y ramas suberectas; hojas ampliamente ovadas o ovado-lanceoladas, ápice cortamente acuminado   *S. subpinnatum*
1. Plantas formando esteras, subpinnadas, tallos secundarios y ramas postradas; hojas ovado-acuminadas, ápice acuminado   *S. subsimplex*

Sematophyllum subpinnatum (Brid.) Britt., Bryologist 21: 28. 1918. (Fig. 39A-B, D). Synonym: *Sematophyllum caespitosum* auct. non. (Hedw.) Mitt.

Napo: El Napo, 500 m, *Benoist 4678* (Thériot, 1936); Mishuallí, ca. junction of Río Mishuallí–Río Napo, 01°02'S, 77°41'W, 500 m, *Holm-Nielsen 19277* (AAU). **Pastaza:** Río Bobonaza, 460 m, *Spruce 992, 993* (Mitten, 1869). Illustrations based on *Holm-Nielsen 19280* (AAU) and *Holm-Nielsen 19277* from Ecuador.

Reported from Chimborazo, Loja, Pichincha, Tungurahua, and also the Galapagos Islands; at elevations from 460–1540 m.

Sematophyllum subsimplex (Hedw.) Mitt., J. Linn. Soc., Bot. 12: 494. 1869. (Fig. 39C).

Napo: Añangu, ca. 75 km east of Coca, ca. 00°32'S, 76°23'W, 245–325 m, *Churchill & Sastre-De Jesús 13782a* (NY; Churchill *et al.*, 1992). **Sucumbíos:** Reserva Faunística Cuyabeno, near Laguna Grande, 00°00', 76°12'W, 265 m, *Balslev 84913* (AAU); *Heikkinen RH-1990-107a & -145* (NY). Illustration based on *Balslev 84913*.

In Ecuador reported also from Morona-Santiago and Tungurahua; at elevations from 245–1845 m.

Taxithelium Spruce ex Mitt.

Taxithelium planum (Brid.) Mitt., J. Linn. Soc., Bot. 12: 496. 1869. (Fig. 40A).

Plants rather small, forming thin to rather dense mats; stems and branches spreading, leaves ±complanate, broadly ovate-oblong, 0.8–1.2 mm long, to 0.6 mm wide, concave, broadly acute, margins plane, serrulate throughout by projecting papillae, costae absent, median cells linear, pluripapillose, papillae ca. 3–6 in a row, alar region differentiated, cells quadrate to short rectangular or oval; autoicous; setae slender, to 2 cm long, capsules strongly curved, exostome striate, papillose distally; opercula conic-mamillate.

This species, the only member of the genus in Ecuador, is common and widespread in the Neotropical wet lowlands.

Napo: Zatayaca, 500 m, *Mexia 7072a* (GB); Añangu, ca. 75 km east of Coca, ca. 76°23'W, 00°32'S, 245–325 m, *Churchill & Sastre-De Jesús 13743b* (NY; Churchill *et al.*, 1992); Coca-Auca road, km 53, 76°52'W, 00°50'S, 400 m, *Holm-Nielsen et al. 19759* (AAU); Río Yasuní, Garza Cocha, 75°45'W, 01°01'S, 330 m, *Holm-Nielsen et al. 19917* (AAU); ca. 6.7 km S of Tena on road to Puyo, 77°48'W, 01°02'S, ca. 600 m, *Churchill & Sastre-De Jésus 13656* (NY).

Pastaza: Río Bobonaza, 395 m, *Spruce 941* (NY; Mitten, 1869); Curaray, 200 m, *Andersson 870* (GB); Finca Valle de Muerte on Río Curaray, 200 m, *Andersson 900* (GB). **Sucumbíos:** Lagunas de Cuyabeno, 00°01'S, 76°11'W, 250 m, *Brandbyge et al. 30516* (AAU); Reserva Faunística Cuyabeno, N of Laguna Grande, 00°01'N, 76°11'W, 265 m, *Heikkinen RH-1990-60* (NY).. Illustration based on *Brandbyge et al. 30516.*

In Ecuador reported also from Guayas, Loja, Los Ríos and Pichincha; at elevations from 30–1220 m.

Trichosteleum Mitt.

Plants forming thin to rather dense mats; stems and branches spreading or somewhat subascending; leaves subcomplanate or not, ovate-lanceolate, often subfalcate-secund, short to long acuminate, margins plane, weakly serrate to papillose-serrate, ecostate, median cells rhombic to linear, unipapillose, papillae indistinct to large and conspicuous, alar cells inflated, few, oval to short rectangular-rounded; autoicous, setae often roughened distally, capsules pendulous, exostome striate, furrowed, endostome basal membrane high, cilia 1–2 or reduced, opercula long rostrate.

The papillae are often difficult to observe under a light microscope, however they can often be observed when the leaves are folded, or on the distal leaves of young branches. The genus is a common element of the wet lowland forests and occasionally extends into the montane forests. At least three species have been recorded for Ecuador.

Key to the Species

1. Leaf apices bluntly acute to rounded-acute; apical cells rhombic to broadly rhomboidal   *T. fluviale*
1. Leaf apices acuminate; apical cells fusiform to rhomboidal   **2.**
   2. Leaves 1–1.2 mm long, propagula absent, median cell papillae distinct   *T. papillosum*
   2. Leaves 0.6–0.8 mm long, propagula produced on rhizoids in axial of branch leaves, median cell papillae indistinct   *T. sp.*

Clave para las Especies

1. Apices de la hoja despuntadamente agudos hasta agudo-redondeados; céllulas apicales rombicas hasta ampliamente romboidales   *T. fluviale*
1. Apices de la hoja acuminados; céllulas apicales fusiformes hasta romboidales   **2.**
   2. Hojas de 1–1.2 mm de longitud, propágulos ausentes, papilas de las células medias bien definidas   *T. papillosum*
   2. Hojas de 0.6–0.8 mm de longitud, propágulos producidos sobre rizoides en los ejes de hojas ramificadas, papilas de las células medias no bien definidas   *T. sp.*

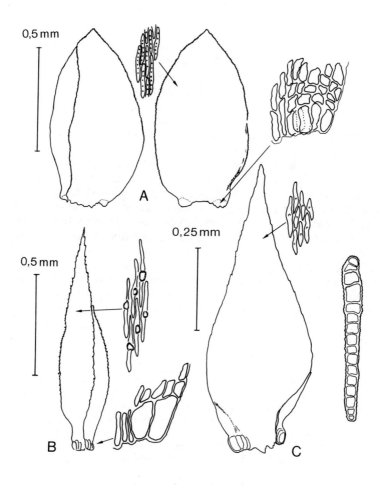

Figure 40. A. *Taxithelium planum*, A. leaves; B. *Trichosteleum papillosum*, B. leaf; C-D. *Trichosteleum sp.*, C. leaf, D. propagula.

Trichosteleum fluviale (Mitt.) Jaeg., Ber. St. Gall. Naturw. Ges.
1875–1876: 419. 1878.

Napo: Río Yasuní, Garza Cocha, 01°01'S, 75°45'W, 330 m, *Holm-Nielsen et al. 19884*
(AAU).

Trichosteleum papillosum (Hornsch.) Jaeg., Ber. St. Gall. Naturw.
Ges. 1875–1876: 419. 1878. (Fig. 40B).
Leaves 1–1.2 mm long, ca. 0.3 mm wide, long acuminate, margin
plane, serrate-papillose throughout except at base, median cell
papillae high and conspicuous; setae ca. 10 mm long, slightly rough
distally, urn ca. 0.5–0.6 mm long.

Napo: Añangu, ca. 75 km east of Coca, ca. 00°32'S, 76°23'W, 245–325 m, *Brako 5339* (NY).
Pastaza: Curary, 200 m, *Andersson 883* p.p. - with *Pilosium* (AAU). **Sucumbíos:** Reserva
Faunística Cuyabeno, near Laguna Grande, 00°00', 76°12'W, 265 m, *Balslev 84914* (AAU), N of
Laguna Grande, *Heikkinen RH-1990-126 & -156* (NY). Illustration based on *Churchill et al.
15484* (NY) from Colombia.

Trichosteleum sp. (Fig. 40C-D).
Although the identity of this species still remains unknown, the
distinctive papillose propagula on rhizoids in the axil of distal
branch leaves is noteworthy, and possibly not reported for the
genus, at least in the Neotropics. Such asexual propagation is
apparently a common feature in the tropics unlike the situation in
the temperate region. William Buck (pers. comm.) suggested that
this collection resembled *Sematophyllum subsimplex*, but having
papillae, possibly an abberent form. At present I prefer to retain it
under the genus *Trichosteleum*.

Sucumbíos: Reserva Faunística Cuyabeno, near Laguna Grande, 00°00', 76°12'W, 265 m,
Balslev 84912 (AAU, NY). Illustration based on *Balslev 84912*.

SPLACHNOBRYACEAE

Splachnobryum C. Müll.

Splachnobryum obtusum (Brid.) C. Müll., Verh. Zool. Bot. Ges.
Wien 19: 504. 1869. (Fig. 41A-B).
Plants small, growing in loose tufts; stems erect, leaves obovate
to oblong, 0.8–1 mm long, to 0.4 mm wide, apex broadly acute to
rounded, margins plane, recurved below, entire to slightly
crenulate distally, costae percurrent or ending several cells below
apex, median cells rhomboidal to hexagonal, smooth; sporophyte
terminal, dioicous; setae elongate, ca. 5 mm long, smooth; capsules
erect, urn ovoid-cylindrical, ca. 1 mm long, peristome single,
papillose; opercula conic; calyptrae cucullate.

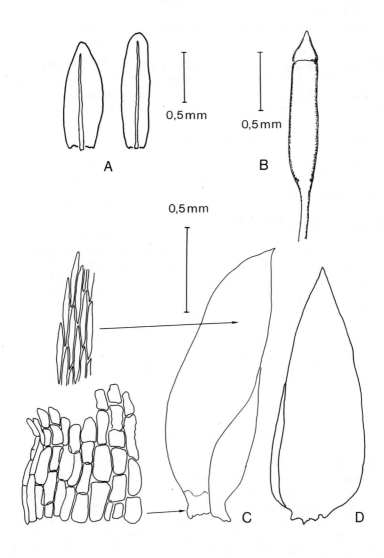

Figure 41. A-B. *Splachnobryum obtusum*, A. leaves, B. capsule; C-D. *Pilosium chlorophyllum*, C. lateral leaf, D. submedian leaf.

Only recently reported for Ecuador, this species is scattered throughout the lowland Neotropics. Our collections are from a disturbed site around a compound, growing on concrete blocks.

Napo: Añangu, ca. 75 km east of Coca, ca. 00°32'S, 76°23'W, 245–325 m, *Churchill & Sastre-De Jesús 13840, 13844* (NY; Churchill *et al.*, 1992). Illustration based on *Churchill & Sastre-De Jesús 13840*.

STEREOPHYLLACE

Pilosium (C. Müll.) Fleisch.

Pilosium chlorophyllum (Hornsch.) C. Müll. *in* Broth., Flora 83: 340. 1897. (Fig. 41C-D).

Plants forming mats; stems spreading, leaves loosely complanate, median and lateral leaves differentiated, lateral leaves ±asymmetric, oblong, 1.5–2 mm long, to 0.7 mm wide, broadly acuminate or acute, margins folded at base, distally plane, entire, ecostate or short and forked, median cells linear, smooth, alar region asymmetrically differentiated, one side with more numerous cells, cells quadrate to rectangular, golden-brown, median leaves symmetric, alar region symmetrically distributed; setae elongate, to 15 mm long; capsule inclined, urn ovoid to short cylindrical, 1–1.2 mm long, peristome double, exostome teeth 16, striate below, endostome segments 16, cilia 1; calyptrae cucullate, smooth.

Napo: Añangu, ca. 75 km east of Coca, ca. 00°32'S, 76°23'W, 245–325 m, *Churchill & Sastre-De Jesús 13859* (AAU, NY; Churchill *et al.*, 1992). **Pastaza:** Curaray, 200 m, *Andersson 883* (AAU); Curaray, northern bank 2 km W of school, 01°22'S, 76°58'W, 250 m, *Holm-Nielsen et al. 21921* (AAU); Lorocachi, along Río Curaray, 01°38'S, 75°58'W, 200 m, *Jaramillo et al. 31391* (AAU). **Sucumbíos:** Reserva Faunística Cuyabeno, near Laguna Grande, 00°00', 76°12'W, 265 m, *Balslev 84911* (AAU), N of Laguna Grande, *Heikkinen RH-1990-71 & -154* (NY). Illustration based on *Andersson 883*.

Widespread in the Neotropics, this species is commonly found on logs and base of trees or treelets. *Pilosium* has traditionally been placed with the Plagiotheciaceae, but has been removed from that family by Buck and Ireland (1985) suggesting that it be placed in the Hookeriaceae s. l. Based on similar gametophytic features, I prefer to place it in the Stereophyllaceae (sensu Buck and Ireland, 1985).

THAMNOBRYACEAE

Plants often dendroid or frondose, primary stems creeping; secondary stems erect, usually stipate, pinnate branched above stipe, flagellate branches frequent, stipe leaves small, often triangular, appressed or spreading-recurved, secondary and branch leaves radially foliate to loosely complanate, ovate, ovate-oblong to ovate-lanceolate, margins distally serrate, costae single, 2/3–3/4 lamina length, upper and median cells usually rhomboid-rounded to broadly fusiform, cells smooth to rarely papillose; dioicous; setae lateral, short or elongate, capsules erect, urn ovoid, peristome double, exostome papillose or striate, endostome basal membrane low, segments papillose; opercula rostrate; calyptrae cucullate, smooth and naked.

See comments under Neckeraceae in reference to the family and genera placed here. Reference: Sastre-de Jesús (1987).

Key to the Genera

1. Plants small, to 2.5 cm tall; leaf apex rounded or obtuse, weakly serrulate   *Pinnatella*
1. Plants large, mostly 4 cm or more tall, leaf apexacute to acuminate, sharply serrate to entire   *Porotrichum*

Clave para los Géneros

1. Plantas pequeñas, hasta 2.5 cm de altura; ápice de la hoja redondeado u obtuso, debilmente serrulado   *Pinnatella*
1. Plantas grandes, la mayoría de 4 cm de altura o más; ápice de la hoja agudo hasta acuminado, fuertemente serrado hasta entero
                                                                        Porotrichum

Pinnatella Fleisch.

Pinnatella minuta (Mitt.) Broth., Nat. Pfl. 1(3): 857. 1906. (Fig. 42A-B).

Plants small, to ca. 2.5 cm tall; secondary stems ±frondose, branches often flagelliform, secondary stem leaves ovate, 0.6–0.8 mm long, 0.3–0.4 mm wide, apex rounded or obtuse, margins weakly serrulate distally, costae ending below the apex, costa apex often forked at, median cells irregularly quadrate-rounded, bulging or appearing mammillose, thick-walled; setae short, roughed or papillose distally; capsules exserted, exostome striate (sporophyte not observe).

Two species of *Pinnatella* have been reported from Ecuador.

Napo: Añangu, ca. 75 km east of Coca, ca. 00°32'S, 76°23'W, 245–325 m, *Churchill & Sastre-De Jesús 13880b* (AAU; Churchill et al., 1992). Illustration based on *Churchill & Sastre-De Jesús 13880b*.

Porotrichum (Brid.) Hampe

Plants medium to rather large; stipate leaves appressed to recurved, secondary stems mostly frondose, branches often flagellate, secondary stem and branch leaves ovate to oblong-lanceolate or -ligulate, margins often strongly serrate distally, costae 2/3 or more lamina length, median cells irregularly rhomboidal to rhomboidal-fusiform, smooth or occasionally weakly papillose, thick-walled; dioicous; setae elongate, smooth; capsule erect or suberect, peristome papillose or striate at base.

The genus *Porothamnium* has been placed within *Porotrichum* by Sastre-De Jesús (1987). Although some 15 species are listed for Ecuador, probably several will prove to be synonyms.

Key to the Species

1. Stipe leaves spreading; leaf apex obtuse-apiculate *P. filiferum*
1. Stipe leaves appressed; leaf apex acute **2.**
 2. Secondary stem leaves broadly ovate, acute, ±entire *P lindigii*
 2. Secondary stem leaves ovate-lanceolate or -ligulate, serrate **3.**
  3. Secondary stem leaves ovate-lanceolate or -oblong, serrate in distal half *P. longirostre*
  3. Secondary stem leaves ovate-lingulate, serrate only at apex *P. mutabile*

Clave para las Especies

1. Pecíolo de las hojas orientado verticalmente con relación al tallo, ápice de la hoja obtuso-apiculado *P. piliferum*
1. Pecíolo de las hojas apresado al tallo; ápice de la hoja agudo **2.**
 2. Hojas del tallo secundario ampliamente ovadas, agudas, ±enteras *P. lindigii*
 2. Hojas del tallo secundario ovado-lanceoladas o -liguladas, serradas **3.**
  3. Hojas en tallos secundarios ovado-lanceoladas u oblongas, serrados en la mitad distal *P. longirostre*
  3. Hojas en tallos secundarios ovado-liguladas, serrados solo hacia el ápice *P. mutabile*

Porotrichum filiferum Mitt., J. Linn. Soc., Bot. 12: 468. 1869. Synonym: *Porothamnium filiferum* (Mitt.) Fleisch.

Pastaza: Río Bobonaza, 460 m, *Spruce 1365* (NY; Mitten, 1869).

In Ecuador reported from Zamora; to 2100 m.

Porotrichum lindigii (Hampe) Mitt., J. Linn. Soc., Bot. 12: 461. 1869. (Fig. 36A-B). Synonym: *Pireella cavifolia* (Card. & Herz.) Card.

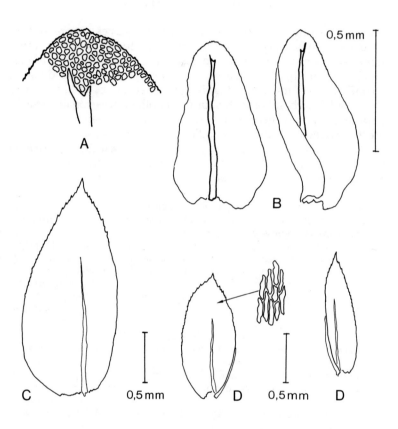

Figure 42. A-B. *Pinnatella minuta,* A. leaf apex, B. leaves; C-D. *Porotrichum longirostre,* C. stem leaf, D. branch leaves.

Napo: Shinguipino, between Río Napo and Tena, ca. 460 m, *Grubb et al. 2934* (Bartram, 1964). Illustration based on *Churchill et al. 14023* (NY) from Colombia.

This species in more recent years has been known under the name *Pireella cavifolia*, fide I. Sastre-De Jesús (pers. comm.).

Porotrichum mutabile Hampe, Flora 45: 456. 1862. Synonym: *Porotrichum variabile* Hampe
Napo: El Napo, 500 m, *Benoist 4665* (Thériot, 1936).

In Ecuador reported also from Morona-Santiago and Pichincha; at elevations from 500–2320 m.

Porotrichum longirostre (Hook.) Mitt., J. Linn. Soc., Bot. 12: 461. 1869. (Fig. 42C-D).
Napo: Shinguipino, between Río Napo and Tena, ca. 460 m, *Grubb et al. 2934a, 2936a* (Bartram, 1964); Añangu, ca. 75 km east of Coca, ca. 00°32'S, 76°23'W, 245–325 m, *Churchill & Sastre-De Jesús 13853* (NY; Churchill et al., 1992). Illustration based on *Andersson 970* (AAU) from Ecuador.

In Ecuador reported also from Chimborazo, Morona-Santiago, Pastaza and Tungurahua; at elevations from 460–2800 m.

THUIDIACEAE

Thuidium B.S.G. *sensu lato*

Plants forming loose to dense mats or wefts; primary stems creeping; secondary stems spreading or occasionally subascending, bi- or tripinnately branched; stem and branches with few to many paraphyllia, papillose; stem and branch leaves similar or dimorphic, stem leaves ovate-lanceolate or triangular, acute to abruptly long acuminate, margins smooth to serrulate-papillose, plane to recurved below, costae single, percurrent to short excurrent, median cells oval, rhombic or irregularly rectangular, thick-walled, uni- or pluripapillose, papillae low to ±long and curved; autoicous or dioicous; sporophytes lateral; setae elongate, smooth to papillose; capsules suberect to horizontal, cylindrical, curved or not, peristome double, exostome teeth 16, cross-striate below, papillose above, endostome membrane high, segments 16, keeled, cilia 2–3; opercula short to long rostrate, oblique; calyptrae cucullate, smooth.

Thuidium is generally found in shaded placed, on trunks of trees and shrubs, and on logs. About 15 species have been recorded for Ecuador, however the actual number may be fewer after careful study of this group. The treatment by Gier (1980) is somewhat useful but did little to actually resolve the problems in the genus.

Thuidium has been divided, recognizing *Cyrtohypnum* (includes *T. involvens* and *T. campanulatum* given below) and *Thuidium* s.s. (*T. tomentosum*); see the checklist in Appendix. References: Buck & Crum (1990); Gier (1980).

Key to the Species

1. Plants ±large; secondary stem and branch leaves differentiated, the latter smaller; papillae of branch leaf cells distinct and curved, 1–3; dioicous   *T. tomentosum*
1. Plants small; secondary stem and branch leaves similar; papillae of branch leaf cells minute, ±3–4 ; autoicous   **2.**
   2. Plants ±pinnately branched   *T. involvens*
   2. Plants tripinnately branched   *T. campanulatum*

Clave para las Especies

1. Plantas ±grandes; hojas de tallos secundarios y ramas diferenciadas, estas últimas más pequeñas; papilas de las células de las hojas en las ramas distintivas y curvadas, 1–3; dioicas   *T. tomentosum*
1. Plantas pequeñas; hojas de tallos secundarios y ramas similares; papilas de las células de las hojas en las ramas diminutas, ±3–4; autoicas   **2.**
   2. Plantas ±pinnadamente ramificadas   *T. involvens*
   2. Plantas tripinnadamente ramificadas   *T. campanulatum*

Thuidium campanulatum Mitt., J. Linn. Soc., Bot. 12: 574. 1869.

Napo: SW of Nuevo Rocafuerte, along Río Braga, 200–230 m, *Jaramillo & Coello 4608* (AAU); Añangu, ca. 75 km east of Coca, ca. 00°32'S, 76°23'W, 245–325 m, *Churchill & Sastre-De Jesús 13743, 13792, 13797* (NY; Churchill *et al.*, 1992).

I previously considered the identity of the above collections from Añangu to represent *T. schistocalyx* (C. Müll.) Mitt., however I believe they better match the name presently used.

Thuidium involvens (Hedw.) Mitt., J. Linn. Soc., Bot. 12: 575. 1869. (Fig. 43A-E).

Morona-Santiago: Patuca, 600 m, *Harling 2278b, 2279* (Crum, 1957). **Napo:** El Napo, 500 m, *Benoist 4678* (Thériot, 1936); km 40 SE of Coca, 250 m, *Fransén 39* (AAU, GB); Añangu, ca. 75 km east of Coca, ca. 00°32'S, 76°23'W, 245–325 m, *Brako 4638b* (NY); Río Suno, 00°42'S, 77°10'W, 400 m, *Holm-Nielsen & Jeppesen 902* (AAU, GB; Robinson *et al.*, 1971). **Sucumbíos:** Reserva Faunística Cuyabeno, N of Laguna Grande, 00°01'N, 76°11'W, 265 m, *Heikkinen RH-1990-310* (NY). Illustration based on *Fransén 39*.

In Ecuador reported also from Guayas; at elevations from 245–700 m.

Thuidium tomentosum Besch., Mém. Soc. Sc. Nat. Cherbourg 16: 237. 1872. Synonym: *Thuidium antillarum* Besch.

Morona-Santiago: Patuca, 600 m, *Harling 2285b, 2290b* (S; Crum, 1957). **Napo:** Shinguipino, between Río Napo and Tena, ca. 460 m, *Grubb et al. 2953* (Bartram, 1964).

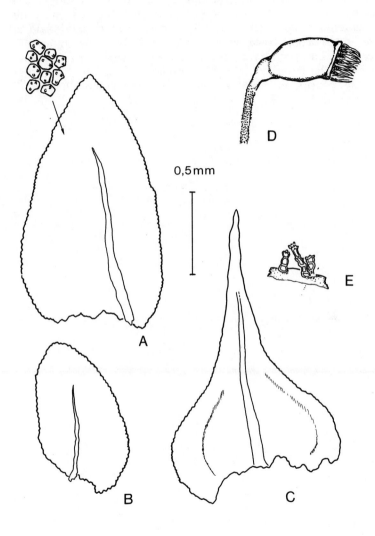

Figure 43. A-E. *Thuidium involvens*, A. secondary stem leaf, B. branch leaf, C. primary stem leaf, D. capsule, E. paraphyllia.

The report of *T. pseudo-delicatulum* Jaeg. given by Crum (1957) appears to match *T. tomentosum*, however the first may be encountered in Amazonas. *Thuidium pseudo-delicatulum* is said to have an excurrent costa, where as *T. tomentosum* Besch. and *T. urceolatum* Lor. both have costa that are percurrent or end below the apex. The latter two are differentiated by the number of the papillae, pluripapillose in *T. tomentosum* and unipapillose in *T. urceolatum*.

6. GLOSSARY

This glossary is intended to cover terms mostly related to mosses, and in general does not cover typical terms that can be found in basic botanical glossaries.

abaxial. Back or dorsal side of the leaf, away from the stem.

acrocarpous. Sporophyte produced on terminal stem or main branch; characteristic of mosses that are erect and grow in tufts or cushions. Figure 44.

alar cells. Region at basal angles of the leaf; cells can be undifferentiated and similar to upper median cells or if differentiated, then in shape and size, often oval, quadrate or rectangular. Figure 46.

amphigastria. Dorsal and/or ventral leaves differentiaed in shape and size from lateral leaves.

bicostate. Leaf with 2 costae (nerves), characteristic of the Callicostaceae or Hypnaceae where the costa is short and forked. Figure 45.

bistratose. Leaf lamina in 2 layers.

bordered. Leaf margin with cells often linear and hyaline, differentiated from the inner lamina cells which often are larger and broader. Figure 46.

calyptra. Gametophytic tissue that envelopes the early developing sporophyte; the calyptra appears as a mitriate or cucullate cap "protecting" the capsule as it matures. Figure 49.

cancellinae. Cells that are generally enlarged and hyaline, mostly confined toward the leaf base; usually strongly differentiated from distal cells, as in the Calymperaceae.

capsule. Spore bearing sturcture; composed of the neck, between the seta and urn, and the urn containing the sporangium. Figure 48.

chlorocysts. Elongate chlorophyll leaf cells enclosed by hyaline (leucocyst) cells, as in the Leucobryaceae.

cilia. Narrow filaments that alternate between the segments on the basal membrane of the inner peristome (endostome). Figure 50.

complanate. Stem and/or branch leaves compressed or flattened.

corticolous. Growing on bark.

costa. Nerve of the leaf, either single (unicostate) or double (bicostate), the latter often appearing short and forked. Figure 45.

costate. With a costa; **ecostate** - without a costa.

cucullate. Type of calyptra that is conic in shape and split on one side. Figure 49.

dendroid. Tree-like growth form, with a trunk-like base (stipe) and branched above. Figure 44.

dioicous. Antheridia and archegonia on separate plants.

doubly serrate. Marginal teeth appearing in pairs (biserrate).

ecostate. Leaf lacking a costa or nerve.

endostome. Inner series of a double peristome, composed of a basal membrane with alternating segments and cilia (cilia in some taxa are reduced or absent). Figure 50.

excurrent. Costa extending beyond the leaf lamina, which can be short or elongate, entire or toothed. Figure 45.

exostome. Outer series of a double peristome, usually consisting of 16 teeth that are entire or variously divided. Figure 50.

flagelliform-branches. Slender branches with reduced leaves, usually found on distal stems or branches; flagelliferous branches appear to be a rather common means of asexual reproduction in tropical mosses.

frondose. Growth form often with a stipe base and distally, often regularly, branched above and in one plane.

gametophyte. Leafy haploid, sexual generation of a moss.

gemmae. Small globose or cylindrical celled structures produced on leaves or in the axil of leaves; a further means of asexual reporduction.

gregarious. Plants growing somewhat separated from each other.

hyaline. Transparent or colorless cells, often in reference to leaf cells such as bordered margins, or to excurrent costa; also referring to the peristome, particularly the endostome.

leucocysts. Usually enlarged colorless leaf cells, often enclosing the chlorophyll cells (chlorocysts) as in the Leucobryaceae.

lignicolous. Growing on decaying wood, usually logs.

limbate. With a border, usually refering to the bordered leaf margin or intramaginal region; **elimbate** - lacking a differentiated border.

mammillose. Central portion of a leaf cell that is bulging, forming a rounded or blunt projection. Cf. papillose. Figure 47.

mitrate. Type of calyptra that is conic, often campanulate, with the base divided or split with few to several lobes. A mitrate calyptra may also be smooth, plicate and/or hairy. Figure 49.

microphyllous branches. Reduced, small branches and leaves, often produced in the axil of distal secondary stem and branch leaves.

These structures likely serve as a means of asexual reproduction.

monoicous. Archegonia and antheridia on the same plant, generally they can be mixed together, synoicous, or the inflorescences can be separated, autoicous.

neck. Region between the seta and urn, absent or indistinct in many mosses, but in some, like *Trematodon* the neck is elongate.

operculum. The lid covering the mouth of the capsule or urn. When the capsule is deoperculate, the spores are allowed to disperse. Figure 48.

papillose. Projecting thickening of the leaf or seta cell walls. Papillae can be single, rounded or sharply pointed, or branched; **unipapillose** - single papillae, **pluripapillose** - 2 or more papillae, projecting from the cell wall. Figure 47.

paraphyllia. Small filamentous or scale-like structures on the stems and branches of the gametophyte; as in *Thuidium* which are branched and minutely papillose.

percurrent. Costa extending to the tip of the leaf but not beyond; **subpercurrent** - costa ending below the apex. Figure 45.

peristome. Single or double series of teeth like appendages at the mouth of the capsule; if in a single series, then with 16 entire or divided teeth, if in a double series (exostome and endostome), then the outer series usually consist of 16 teeth, the inner series consist of a low to high basal membrane with 16 segments attached to it, often alternating with 1–3 cilia between each segment. Figure 50.

pleurocarpous. Leafy plants producing the sporophyte laterally; usually the gametophytes are prostrate, pendent or dendroid. Figure 44.

propagulum. Asexual gametophytic structure that is often cylindrical or rounded from a stalk that represents an arrested bud, leaf or sometimes a branch.

ramicolous. Growing on branches of trees, treelets or shrubs.

rhizoids. Filaments often serving to anchor the plant to the substrate, found at the base of erect mosses, or on the underside of prostrate mosses, occasionally at the distal tips of branches or sometimes densely covering the stem and forming a tomentum.

rostrate. A beak, referring to the operculum which can be gradually or abruptly narrowed to a short or long point. Figure 48.

saxicolous. Growing on rocks.

secund. Leaves that are turned in one direction on the stem or branch.

seta. Stalk of the capsule; can be short or elongate, smooth, papillose or spinose.

sporophyte. Leafless diploid asexual generation of a moss; consisting of a seta (stalk) and capsule (bearing the spore-sac).

stipe. Trunk-like structure arising from a primary stem; usually the leaves on the stipe are differentiated from those found on the distal secondary stem and branch leaves.

synoicous. Antheridia and archegonia mixed in the same inflorescences; a type of monoicous condition.

teeth. Referring to the peristome; single peristome or the outer series of a double peristome consist of 16 teeth. Also referring to projecting cells along the leaf margin or costa back.

teniola. Intramarginal border, usually of hyaline linear cells, as in the *Calymperes*.

terricolous. Growing on the ground.

tomentose. Referring to a dense mat of usually rusty-red rhizoids.

urn. Spore-bearing structure of the capsule.

wefts. Growth form that is loosely interwoven, with primary and secondary stems prostrate, pendulous or ascending stems.

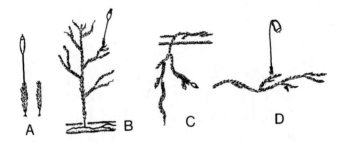

Figure 44. A-B. Growth forms, A. turf form; B. dendroid form; C. pendulous form; D. mat form. A. acrocarpous habit with terminal sporophyte; B-D. pleurocarpous habit with sporophyte lateral.

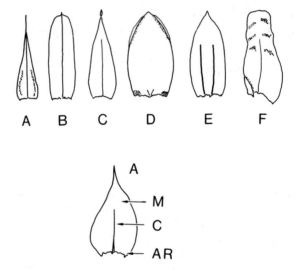

Figure 45. A-G. Leaf shapes and features, A. linear, plicate, apex narrowly acuminate, costa long excurrent, B. ligulate, apex obtuse/rounded, costa mucronate; C. lanceolate, apex acuminate and twisted, costa 3/4 lamina length; D. ovate, concave, apex acute, costa short and forked, alar region differentiated; E. oblong-lanceolate, apex short acuminate, costa long and double; F. ovate-oblong, asymmetrical, surface undulate, apex truncate. G. Leaf features. A. terminology of a typical leaf : A = apex, M = median cell region, C = costa, AR = alar region.

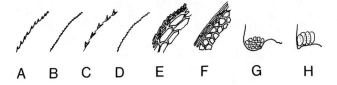

Figure 46. Leaf margin, A. serrate; B. serrulate; C. doubly toothed; D. dentate; E-F. leaves with border, marginal cells strongly differentiated from inner lamina cells; G-H. alar region, G. alar cells quadrate, H. oval and inflated.

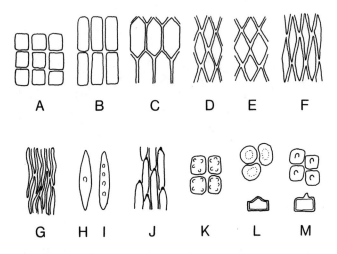

Figure 47. Leaf cell patterns (A-G) and ornamentation (H-M). A. quadrate, B. rectangular, C. hexagonal, D. rhomboidal, E. rhombod, F. fusiform, G. linear, vermicular, H. unipapillose; I. pluripapillose, cells in a row, J. projecting cell angles (unipapillose), K. pluripapillose, L. mammilose, M. papillose.

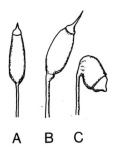

Figure 48. Sporophyte features, A. capsule erect, urn cylindrical, operculum conic short rostrate, B. capsule inclined (suberect), urn ovoid, operculum long rostrate; C. capsule pendulous, urn short ovoid, operculum conic-mammilate.

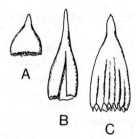

Figure 49. Calyptrae types. A. mitrate, smooth, base erose; B. cucullate, smooth, base entire, C. mitrate-campanulate, plicate, base lobed.

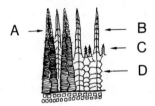

Figure 50. Features of a double peristome. A. exostome, B. endostome segment, C. cilia, D. basal membrane.

7. GLOSARIO

Este glosario está basado en Sastre-De Jesús y Churchill (1986).

abaxial. Lado inferior o dorsal de la hoja.

acrocárpico. Musgos que producen el esporófito sobre los tallos terminales o sobre la rama principal. Término opuesto a pleurocárpico. Característico de los musgos con crecimiento erecto o en forma de racimos y cojines.

alares, células. Región hacia los ángulos basales de la hoja. Estas células pueden estar no diferenciadas y ser similares a las células de la parte media superior, o pueden estar diferenciadas tanto en la forma como el tamaño, a menudo ovaladas, cuadradas o rectangulares.

apiculado. Terminando en una puntita.

arista. Apice terminando en una punta prolongada, usualmente cuando el nervio se extiende fuera de la lámina.

aurícula. Márgen basal de la hoja en forma de lóbulo, a modo de oreja.

autoico. En briófitos, arquegonio y anteridio en diferente inflorescencia en la misma planta.

bicostado. Hoja con dos costas (nervios), característico de las Callicostaceae o Hypnaceae donde la costa es corta y ramificada.

cancelino (na). Células generalmente grandes, y translúcidas ensumayoria ubicadas hacia la, base de las hojas, fuertemente diferenciadas de las células distales, como en Calymperaceae (Fig.).

cápsula. Estructura que encierra las esporas, compuesta por el cuello, entre la seta y la urna, y la urna contiene el esporangio.

cespitoso. Crecimiento en forma de césped.

costa. Nervio de la hoja, puede ser simple (unicostado), o doble (bicostado).

cilios. Estructuras delicadas en forma de hilos entre los segmentos del endóstoma de algunos musgos; también se aplica a los flecos de algunas cofias.

clorocistes. Células clorofilíferas pequeñas y alargadas que forman una red que envuelve los leucocistes en las hojas de Leucobryaceae.

cofia. Cubierta o capucha de forma cónica en la punta de la cápsula; se forma de la mitad superior del arquegonio.

columela. Eje central en la cápsula, alrededor de la cual se desarrollan las esporas.

complanado. Hojas del tallo y/o ramas comprimidas en un mismo plano o aplanadas.

conduplicado. Doblado a lo largo de su nervio medio.

corteza. En musgos las capas externas diferenciadas del tallo o ramas.

cuculado. Tipo de caliptra en forma cónica con una hendidura a un lado.

cuello. Parte inferior de la cápsula formado de tejido estéril que puede tomar formas variadas.

decurrente. Con los márgenes de la base de las hojas extendiendose más abajo del punto de inserción de la hoja.

deflexo. Curvado hacia abajo.

dendroide. Crece con ramificaciones similares a las de un árbol, con base similar a un tronco y ramificado hacia arriba, en varios planos.

dentado. Con prominencias a modo de dientes dirigidas hacia afuera.

dientes. Divisiones del perístoma.

dioico. En briófitos, arquegonios y anteridios en diferentes plantas.

dístico. Arreglo en dos filas.

endóstoma. Serie interior de un perístoma doble, compuesto de una membrana basal frecuentemente alternando con cilios.

esporófito. En plantas con alternancia de generaciones, la generación que presenta las esporas.

estípite. Tallo no ramificado de los musgos dendroides.

excurrente. Costa extendiendose más alla de la lámina de la hoja, la cual puede ser corta o elongada, lisa o dentada.

exóstoma. Serie externa de un perístoma doble, usualmente consiste de 16 dientes enteros o divididos.

falciforme. En forma de hoz.

fasciculado. Agrupado, formando manojos que proceden de un mismo punto, como en *Sphagnum* y *Philonotis*.

filiforme. En forma de hebra, delgado y alargado.

flagelado. En briófitos se refiere a ramas atenuadas y largas que parecen flagelos.

frondoso. Crecimiento con una base a forma de tronco y distalmente ramificado en forma regular, pero en un solo plano.

giboso. Con una pequeña joroba, con un lado protuberante.

globoso. Esférico.

hialino. Transparente, sin color.

hialociste. En *Sphagnum* o Leucobryaceae células sin color que acumulan agua y se encuentran rodeadas por clorocistes.

homomalo. Dícese de las hojas tornadas en la misma dirección.

imbricado. Hojas que se cubren, dispuestas como en un tejar.

inclinadas. Se refiere a la posición de las cápsulas, pendular.

incurvado. Se aplica a los márgenes o ápices que se encorvan hacia arribao hacia adentro.

indehiscente. En musgos se aplica a las cápsulas que no poseen opérculo.

lamela. Bordes o crestas en la costa o lámina de algunos musgos.

lanceolada. En forma de laza.

leucocistes. Usualmente células agrandadas de la hoja, descoloreadas (sin clorofila) que se encuentran rodeadas de clorocistes como en Leucobryaceae.

lignícola. Que vive o se desarrolla sobre madera.

lumen. Cavidad celular.

mamiloso. Con protuberancias que parecen mamas.

mitrado. Tipo de caliptra en forma cónica, a menudo campanulada con la base dividida o con una hendidura de pocos o varios lóbulos. Una caliptra mitrada puede ser también lisa, plicada y/o peluda.

mucrón. Punta corta y abrupta en el extremo de cualquier órgano.

muriculado. Con pequeñas púas.

nervio. Se refiere a la nervadura central en las hojas de muchos musgos. Binervio, uninervio.

nudoso. Aplícase a las células con paredes sinuosas.

oblongo. Más largo que ancho.

opérculo. Tapa que cubre la boca de la cápsula o urna.

papiloso. Con protuberancias sobre el lumen o en el extremo de las células, en forma de papila.

parafilo. Pequeñas estructuras desde foliares a filiformes en el tallo de musgos pleurocárpicos.

paráfisis. Filamentos entre los anteridios y arquegonios, de color amarilloso o hialino.

patente. Formando un ángulo de 45° o más, áplicase a la posición de las hojas con relación al tallo.

periquecio. Inflorescencia femenina formada por hojas y arquegonios.

perístoma. Serie de dientes simple o doble, similares a apéndices en la boca de la cápsula.

pleurocárpico. Musgos que producen el esporófito lateralmente, usualmente los gametófitos son postrados, pendientes o dendroides.

plegado. Con dobleces longitudinales.

poros. Pequeñas aperturas o fosas en las paredes de algunas células.

propágulos. Estructuras gametofíticas asexuales, cilíndricas o redondeadas en las hojas o ramas reducidas.

protonema. Filamentos producidos a partir de la germinación de una espora, da origen al gametófito.

recurvado. Encorvado hacia abajo y hacia atrás, se apica a los márgenes de las hojas, dientes, y perístoma.

rizoides. Filamentos simples o ramificados que sirven a la planta de anclaje al substrato, se encuentran hacia la base de musgos erectos o sobre la superficie inferior de los musgos postrados, ocasionalmente pueden ser encontrados al extremo de las ramas o algunas veces cubriendo densamente el tallo formando un tomento.

rostrado (a). Picudo, que termina en un pico.

rugulado (da). Con pliegues transversales u ondulado.

saxícola. Que crece sobre rocas.

segmentos. Las divisiones en forma de dientes del perístoma interno.

serrado. Con dientes marginales apuntando hacia afuera.

sésil. Sin seta, por lo tanto la cápsula queda entre las hojas del periquecio.

seta. Pedúnculo que sostiene la cápsula, puede ser corto o elongado, liso, papiloso o espinoso.

teniola. Hilera de células rectangulares y hialinas intramarginalmente.

truncado. Apice abruptamente cuadrado.

unicostado. Hoja con una sola costa (nervio).

urna. La parte de la cápsula que encierra las esporas.

vaina. Se aplica a la parte conduplicada en las hojas de *Fissidens*.

8. APPENDIX

A CHECKLIST OF THE MOSSES OF ECUADOR

The following list of the mosses of Ecuador, including the Galapagos Islands, is based on the literature review given by Steere (1948) and publications since that time (Bartram, 1955, 1964; Churchill *et al.* 1992; Crum, 1957: Gradstein & Weber, 1982; Robinson *et al.*, 1971, 1977), including revisionary studies. Names of species follow Index Muscorum (Wijk *et al.*, 1959–1969) and recent taxonomic studies available to the author. As discussed in the introduction, the late Dr. William Campbell Steere had prepared a detailed annotated catalogue of the Ecuadorean mosses. That work will include considerable information including many new additions to Ecuador, synonyms, citation of specimens by province, and literature.

One motive for providing this checklist in the AAU Reports series is that it has a wider access to those in Ecuador compared to various bryological journals which are not readily avialable.

This present effort simply provides a species list and recent synonyms as they pertain to changes since Steere's important 1948 catalogue. Here families, genera and species are listed alphabetically; synonyms are given following the accepted name. References are given at the beginning of each family. Included references have the objective of listing those publications dealing with nomenclatural and taxonomic changes at the species, generic and family level. Furthermore, references are included that could possibly assist in the identification of Ecuadorean mosses, or provide information on the classification of listed taxa, though not followed in all cases in this checklist. Although no attempt is made in this treatment to list all references pertaining to Ecuador, a few references not mentioned previously but of possible interest are the following: Bescherelle (1894), Brotherus (1920), Hampe (1869), Lorentz (1868), Mitten (1851), Spruce (1861), and Taylor (1846, 1847, 1848a,b). Several recent papers dealing with ecology that consider mosses in Ecuador include the following: Løjtnant and Molau (1982), Balslev and de Vries (1982, 1991), and Muñoz *et al.* (1985). Finally, though there are presently no keys to Ecuadorean genera of mosses, the keys by Griffin for Merida, Venezuela (1982), and for Costa Rica (1983) should prove useful for the identification of many Ecuadorean genera. A synoptical treatment of the Colombian mosses is in preparation that should likewise be useful (Churchill & Linares, in prep.).

Based on this list, there are 874 species distributed among 233 genera, and 62 families (Table 4). The ten largest families account for ca. 65% of the total number of mosses recorded for Ecuador (Table 3). The ten largest genera are also given in Table 3.

Table 3. The ten largest families and genera of Ecuadorean mosses.

	Number of species
Families	
Pottiaceae	104
Callicostaceae	93
Dicranaceae	79
Bryaceae	73
Orthotrichaceae	72
Bartramiaceae	40
Hypnaceae	39
Fissidentaceae	27
Genera	
Macromitrium (Orthotrichaceae)	34
Lepidopilium (Callicostaceae)	31
Camyplopus (Dicranceae)	30
Fissidens (Fissidentaceae)	27
Bryum (Bryaceae)	21
Cyclodictyon (Callicostaceae)	18
Barbula (Pottiaceae)	14
Breutelia (Bartramiaceae)	14
Philonotis (Bartramiaceae)	14
Sphagnum (Sphagnaceae)	14

A fair number of the names given in the present checklist will likely represent synonyms after critical taxonomic studies, particularly so for such genera as *Barbula*, *Cyclodictyon*, *Macromitrium*, *Mittenothamnium*, *Philonotis*, and *Tortula*, to name just a few. However, the number of additional records that are likely to be found in the years ahead will increase the number of species known for Ecuador. Thus it seems reasonable that the actual number of mosses for Ecuador will be between 800 to 900 species.

Table 4. Numbers of genera and species for the 62 families of Ecuadorean mosses.

Family	Number of	
	Genera	Species
Adelotheciaceae	1	1
Amblystegiaceae	5	13
Andreaeaceae	2	10
Aulacomniaceae	1	1
Bartramiaceae	7	40
Brachytheciaceae	11	26
Bruchiaceae	2	2
Bryaceae	12	73
Callicostaceae	16	93
Calymperaceae	2	16
Catagoniaceae	1	1
Cryphaeaceae	2	11
Daltoniaceae	3	14
Dicranaceae	18	77
Ditrichaceae	3	7
Encalyptaceae	1	2
Entodontaceae	3	10
Erpodiaceae	1	2
Eustichiaceae	1	1
Fabroniaceae	1	2
Fissidentaceae	1	27
Funariaceae	3	12
Grimmiaceae	4	17
Hedwigiaceae	3	6
Hookeriaceae	1	1
Hydropogonaceae	1	1
Hylocomiaceae	1	1
Hypnaceae	12	29
Hypopterygiaceae	1	2
Leptodontaceae	1	1
Lepyrodontaceae	1	1
Leskeaceae	1	3
Leucobryaceae	3	9
Leucomiaceae	2	2
Meteoriaceae	8	24
Mniaceae	1	1
Myriniaceae	1	3
Neckeraceae	3	11
Orthotrichaceae	7	72
Phyllodrepaniaceae	1	1
Phyllogoniaceae	1	2
Plagiotheciaceae	1	3
Polytrichaceae	6	18
Pottiaceae	33	104
Prionodontaceae	1	5
Pterobryaceae	7	14
Racopilaceae	1	2
Rhacocarpaceae	1	2

Rhegmatodontaceae	1	1
Rhizogonaceae	3	4
Rigodiaceae	1	1
Seligeriaceae	1	1
Sematophyllaceae	8	24
Sorapillaceae	1	1
Sphagnaceae	1	14
Splachnaceae	2	5
Splachnobryaceae	1	1
Stereophyllaceae	4	8
Symphyodontaceae	1	1
Thamnobryaceae	4	20
Thuidiaceae	3	16
Trachypodaceae	1	1
Total	**233**	**874**

Adelotheciaceae

[Reference: Ochyra et al. 1992. Note: *Adelothecium* has previously been placed in the Hookeriaceae *s. l.*, see comments under Callicostaceae.]

Adelothecium
  bogotense (Hampe) Mitt.

Amblystegiaceae

Amblystegium
  fluviatile (Hedw.) B.S.G. (*Hygroamblystegium*)
  serpens (Hedw.) B.S.G.

Campylium
  praegracile (Mitt.) Broth.
  radicale (P. Beauv.) Grout
  trichocladum (Tayl.) Broth.

Cratoneuron
  filicinum (Hedw.) Spruce

Drepanocladus
  capillaceus (Mitt.) Broth.
  exannulatus (B.S.G.) Warnst.
  fluitans (Hedw.) Warnst.
  leitensis (Mitt.) Broth.
  sendtneri (Schimp.) Warnst.
  uncinatus (Hedw.) Warnst.

Hygrohypnum
  pelichuense Williams

Andreaeaceae

Acroschisma
  wilsonii (Hook. f. & Wils.) Lindl. (*Andreaea*)

Andreaea
  brevipes Spruce
  karsteniana C. Müll.
  rupestris Hedw.
  seriata Roth
  squarrosa Mitt.
  striata Mitt.
  subenervis Hook. f. & Wils.
  urophylla H. Robins.

vulcanica Lor.

Aulacomniaceae

[Note: The genus *Leptotheca* is placed in the Rhizogoniaceae.]

Aulacomnium
  palustre (Hedw.) Schwaegr.

Bartramiaceae

[References: Crum and Griffin (1981); Fransén (1988); Griffin (1984a, 1984b, 1988, 1989a); Griffin and Buck (1989); Robinson (1967).]

Anacolia
  laevisphaera (Tayl.) Flow.

Bartramia
  angustifolia Mitt.
  flavicans Mitt.
  humilis Mitt. (*Conostomum jamesonii* (Tayl.) Steere)
  mathewsii Mitt.
  potosica Mont. (*B. thrausta* Schimp. ex Britt., fide Fransén)

Breutelia
  aciphylla (Wils.) Jaeg.
  brachyphylla Broth.
  brittoniae Ren & Card.
  chrysea (C. Müll.) Jaeg.
  inclinata (Hampe & Lor.) Jaeg.
  integrifolia (Tayl.) Jaeg.
  jamaicensis (Mitt.) Jaeg. (*Philonotis*)
  karsteniana (C. Müll.) Jaeg.
  polygastrica (C. Müll.) Broth.
  reclinata Broth.
  scariosula (C. Müll.) Broth.
  subarcuata (C. Müll.) Schimp.
  tomentosa (Brid.) Jaeg. (*B. macrotheca* (Hampe) Jaeg.)
  trianae (Hampe) Jaeg. (*B. allionii* Broth.)

Conostomum
  pentastichum (Brid.) Lindb. (*C. australe* Sw.)
  speirostichum C. Müll.

Flowersia
  setifolia (Hook. & Arnott) Griffin & Buck (*Anacolia, Bartramidula, Leiomela*)

Leiomela
aristifolia (Jaeg.) Wijk & Marg. (*L. aristata* Broth.)
bartramioides (Hook.) Par.
ecuadorensis H. Robins.

Philonotis
andina (Mitt.) Jaeg. (*Hygroamblystegium ecuadorense* Bartr.)
curvata (Hampe) Jaeg.
elongata (Dism.) Crum & Steere (*P. sphaerocarpa* var. *elongata*
Dism.)
erecta (Mitt.) Griffin & Buck (*Bartramidula*)
fontanella (Hampe) Jaeg.
glaucescens (Hornsch.) Broth. (*P. tenella* (C. Müll.) Jaeg.)
gracilenta (Hampe) Jaeg.
gracillima Ångstr.
incana (Tayl.) H. Robins. (*Breutelia*)
osculatiana De Not.
rufiflora (Hornsch.) Reichdt. (*P. elegantula* (Tayl.) Jaeg.)
scabrifolia (Hook. f. & Wils.) Braithw.
sphaerocarpa (Hedw.) Brid.
uncinata (Schwaegr.) Brid.

Brachytheciaceae

[References: McFarland (1988); Menzel (1991); Robinson (1967, 1987).]

Aerolindigia
capillacea (Hornsch.) Menzel (*L. aciculata* (Tayl.) Hampe, *L. densiretis*
Hampe & Lor., *Rhynchostegiella attenuata* Bartr.)

Brachythecium
austro-glareosum (C. Müll.) Kindb.
chocayae Herz.
conostomum (Tayl.) Jaeg.
occidentale (Hampe) Jaeg. (*B. prasophyllum* (Hampe) Jaeg.)
plumosum (Hedw.) B.S.G. (*B. asperulum* (Hampe) Jaeg.)
praelongum Schimp. ex C. Müll. (*B. herzogii* Broth.)
pseudo-piliferum (Hampe) Jaeg.
rutabulum (Hedw.) B.S.G. (*B. pseudo-rutabulum* (Hampe) Jaeg., *B.*
rutabulum var. *columbicum* De Not.)
stereopoma (Mitt.) Jaeg. (*B. alboflavens* Card.)

Eurhynchium
pulchellum (Hedw.) Jenn.

Homalothecium
aequatoriense Thér.

Kindbergia
[*Brachythecium immersum* Thér. to be transfered by McFarland]

Oxyrrhynchium
praelongum (Hedw.) Warnst.
remotifolium (Grev.) Broth. (*Rhynchostegium*)

Palamocladium
leskeoides (Hook.) H. Robins. (*Pleuropus*)

Platyhypnidium
aquaticum (Sauerb.) Fleisch.

Rhynchostegium
campylocarpum (C. Müll.) De Not. (*Eurhynchium*)
conchophyllum (Tayl.) Jaeg.
inerme (Mitt.) Jaeg.
lamasicum (Mitt.) Besch.
parvulum Broth.
ulicon (Tayl.) Jaeg.

Steerecleus
scariosus (Tayl.) H. Robins. (*Rhynchostegium, Eurchynchiella*)
serrulatum (Hedw.) H. Robins. (*Rhynchostegium*)

Stenocarpidiopsis
salicicola (Mitt.) Fleisch. in Broth.

Bruchiaceae

[Reference: Buck (1979). Note: *Trematodon* was previously placed in the Dicranceae.]

Eobruchia
ecuatoriana Steere

Trematodon
humilis Mitt.

Bryaceae

[References: Mohamed (1979); Ochi (1980, 1981); Ochyra (1991); Shaw (1982, 1984, 1985, 1987); Shaw and Crum (1982). Note: *Mielichhoferia* has been replaced by the name *Schizymenium*, and most species listed under *Hapoldontium* are probably *Schizymenium*; not all *Mielichhoferia* have been transferred.]

Acidodontium
exaltatum (Mitt.) Jaeg. (*Bryum; B. manabiae* C. Müll.)
grubbii Bartr.

heteroneuron (Mitt.) Broth.
megalocarpum (Hook.) Ren. & Card.
ramicola (Mitt.) Jaeg. (*Bryum*)
seminerve Hook. f. & Wils.
sprucei (Mitt.) Jaeg.
subrotundum (Tayl.) Hook. f. & Wils.
trachyticola (C. Müll.) Broth.

Anomobryum
conicum (Hornsch.) Broth.
filiforme (Dicks.) Solms in Rabehn. (*A. julaceum* (Brid.) Schimp.)
filinerve (Mitt.) Broth.
orbiculatum (Mitt.) Broth.
prostratum (C. Müll.) Besch.
semiovatum (Brid.) Jaeg.

Brachymenium
acuminatum Harv. in Hook. (*Bryum pungens* Tayl.)
consimile (Mitt.) Jaeg.
exile (Dozy & Molk.) Bosch & Lac. (*B. subsmaragdinum* (C. Müll.) Jaeg.)
fabronioides (C. Müll.) Par. (*Bryum sericeum* Mitt.)
globosum Jaeg. (*Bryum; B. bulbiferum* Bartr., ?*Osculatia colombicum* (De Not.) Broth.)
klotzschii (Schwaegr.) Par. (*B. macrocarpum* Card.)
speciosum (Hook. f. & Wils.) Steere (*B. jamesonii* Tayl., *B. krausei* Hampe & Lor.)
systylium (C. Muell.) Jaeg. (*B. imbricatifolium* (C. Müll.) Steere; *Bryum crinitum* Mitt., *B. lozanoi* C. Müll.)

Bryum
algovicum Sendtn. ex C. Müll.
andicola Hook. in Kunth (*B. billardieri* Schwaegr. of reports)
antisanense Bartr.
apiculatum Schwaegr. (*B. allionii* Broth., *B. crugeri* Hampe in C. Müll.)
argenteum Hedw. (*B. brachy-phyllum* Mitt., *B. candicans* Tayl.)
atrovirens Brid. (*B. erythrocarpum* Schwaegr.)
caespiticium Hedw. (*B. congestum* Mitt., *B. subpilosum* Mitt.)
capillare Hedw. (*B. erythroneuron* Mitt.)
clavatum (Schimp.) C. Müll. (*B. concavum* Mitt.)
coronatum Schwaegr. (*B. gracilifolium* C. Müll.)
densifolium Brid. (*B. linearifolium* C. Müll.)
dichotomum Hedw.
limbatum C. Müll. (*B. maynense* Spruce ex Mitt., *B. socorrense* (Hampe) Mitt.)
osculatianum De Not.
pallens Sw.
pallescens Schleich. ex Schwaegr. (*B. flexisetum* Mitt.)
pseudotriquetrum (Hedw.) Gaertn., Meyer & Schreb.

renauldii Roell. ex Ren. & Card. (*Hygrohypnum ellipticum* Thér.)
soboliferum Tayl.
turbinatum (Hedw.) Turn.

Epipterygium
immarginatum Mitt.

Haplodontium
argentifolium (Mitt.) Jaeg.
diplodontum (Mitt.) Jaeg.
megalocarpum (Arnott) Jaeg. (*H. jamesonii* (Tayl.) Hampe)

Leptobryum
pyriforme (Hedw.) Wils.
wilsonii (Mitt.) Broth.

Mielichhoferia
antisanensis (Bartr.) H. Robins.
laxiretis Thér.
longiseta C. Müll.

Mniobryum
bracteatum Bartr.

Orthodontium
gracile B.S.G. (*Stableria osculatiana* (De Not.) Broth., *S. tenella* (Mitt.) Broth.)
pellucens (Hook.) B.S.G. (*O. confine* Hampe)

Pohlia
cruda (Hedw.) Lindb.
elongata Hedw.
fusifera (Mitt.) Broth.
nutans (Hedw.) Lindb.
papillosa (Jaeg.) Broth.
pluriseta Herz.
wahlenbergii (Web. & Mohr) Andr. (*Mniobryum*)

Rhodobryum
beyrichianum (Hornsch.) Schimp. in Hampe
grandifolium (Tayl.) Schimp. in Par. (*R. rhodocephalum* C. Müll. in Lor.)
procerum (Besch.) Par. (*Bryum caulifolium* C. Müll., *B. procerum* Schimp.)
roseum (Hedw.) Limpr.

Schizymenium
andinum (Sull.) Shaw

campylocarpum (Hook. & Arnott ex Hook.) Shaw (*Mielichhoferia*)
lindigii (Hampe) Shaw (*Mielichhoferia*)
nanum Tayl. (*Mielichhoferia*)
pseudopohlia Shaw

Callicostaceae

[References: Allen (1986, 1990); Allen and Crosby (1986a); Allen *et al.* (1985); Buck, (1987a); Churchill (1988b, 1992); Crosby (1969, 1978); Crosby *et al.* (1985); Welch, (1976). Note: The Hookeriales have been redivided (Buck, 1987a) with five recognized families (Adelotheciaceae, Callicostaceae, Daltoniaceae, Hookeriaceae and Leucomiaceae). This classification differs from both the previous traditional arrangement (cf. Welch, 1976), and more recent classifications by Miller (1971) and Crosby (1974). It seems a reasonable system to use at present. A further development has been the division of the genus *Hookeriopsis* into several genera (not all names considered or transfered); in Ecuador this includes: *Brymela* and *Trachyxiphium*, and an elaboration of *Thamniopsis* (see Buck, 1987a).]

Actinodontium
    sprucei (Mitt.) Jaeg.

Amblytropis
    setosa (Mitt.) Broth.

Brymela
    acuminata (Mitt.) Jaeg. (*Hookeriopsis*)
    cuspidata (Jaeg.) Buck (*Hookeriopsis*)
    rugulosa (Mitt.) Buck (*Hookeriopsis*)

Callicosta
    armata (Broth.) Crosby (*Pilotrichum*)
    bipinnata (Schwaegr.) C. Müll. (*Pilotrichum*)
    fendleri (C. Müll.) Crosby (*Pilotrichum*)
    longicaulis (Broth.) Crosby (*Pilotrichum*)

Callicostella
    aspera (Mitt.) Buck. (*Schizomitrium*)
    merkelii (Hornsch.) Jaeg. (*Schizomitrium*)
    pallida (Hornsch.) Ångstr. (*Schizomitrium*)
    rivularis (Mitt.) Jaeg. (*Schizomitrium*)
    rufescens (Mitt.) Jaeg. (*Schizomitrium*)
    saxatilis (Mitt.) Jaeg.
    strumulosa (Hampe & Lor.) Jaeg.
    subsecunda (Mitt.) Churchill *comb. nov. Hookeria subsecunda* Mitt., J. Linn. Soc., Bot. 12: 358. 1869.

Crossomitrium
    epiphyllum (Mitt.) C. Müll. (*C. orbiculatum* C. Müll.)
    patrisiae (Brid.) C. Müll. (*C. spruceanum* C. Müll.)
    saprophilum Broth.

sintenisii C. Müll. (*C. splendens* Broth.)

Cyclodictyon
aeruginosum (C. Müll.) O. Kuntze
albicans (Hedw.) O. Kuntze
allionii Broth.
benoistii Thér.
bombonasicum (Mitt.) O. Kuntze
caespitosum (Mitt.) O. Kuntze
capillatum (Mitt.) O. Kuntze
castaneum (Mitt.) O. Kuntze
chimborazense (Mitt.) O. Kuntze
cuspidatum O. Kuntze
humile (Mitt.) Broth.
krauseanum (Hampe & Lor.) O. Kuntze
latifolium O. Kuntze
mittenii (Jaeg.) O. Kuntze (*C. tenellum* Broth. hom. illeg.)
obscurifolium (Mitt.) O. Kuntze
roridum (Hampe) O. Kuntze
shillicatense (Mitt.) O. Kuntze

Helicoblepharum
fuscidulum (Mitt.) Broth.
venustum (Tayl.) Broth.

Hemiragis
aurea (Brid.) Kindb.

Hookeriopsis
armata Broth.
cavifolia (Mitt.) Jaeg.
cuspidatissima (Hampe) Broth.
exigua (Mitt.) Jaeg.
subscabrella Fleisch. ex Broth.

Hypnella
diversifolia (Mitt.) Jaeg. (*Neohypnella*)
pallescens (Hook.) Jaeg.
pilifera (Hook. & Wils.) Jaeg. (*H. glandulifera* (Hampe) Jaeg.)

Lepidopilum
acutum Mitt.
affine C. Müll. (*L. allionii* Broth., *L. antisanense* Bartr., *L. mittenii*
    C. Müll., *L. pumilum* Mitt., *L. undulatum* Hampe & Lor. hom. illeg.)
ancepes Mitt.
arcuatum Mitt.
argutidens Broth.
armatum Mitt.

brevifolium Mitt.
brevipes Mitt. (L. gracile Mitt., L. nudum Mitt., L. subgracile Broth.)
caviusculum Mitt.
chloroneuron (Tayl.) Hampe & Lor.
cubense (Sull.) Mitt. (L. semilaeve Mitt.)
curvifolium Mitt.
cuspidans Mitt.
erectiusculum (Tayl.) Mitt.
grevilleanum (Tayl.) Mitt. (L. pulcherrimum Steere)
inflexum Mitt.
intermedium (C. Müll.) Mitt.
krauseanum C. Müll.
leucomioides Broth.
longifolium Hampe (L. integerrimum Mitt., L. robustum Mitt.)
muelleri (Hampe) Mitt.
pallido-nitens (C. Müll.) Par. (L. attenuatum Bartr.)
pectinatum Mitt.
phyllophyllum Broth.
polytrichoides (Hedw.) Brid.
radicale Mitt.
scabrisetum (Schwaegr.) Steere
stillicidiorum Mitt.
surinamense C. Müll. (L. flexifolium (C. Müll.) Mitt.)
tenuifolium Mitt.
tortifolium Mitt. (L. crispifolium Bartr.; Cyclodictyon riparium (Mitt.)
    O. Kuntze)

Pilotrichidium
    callicostatum (C. Müll.) Jaeg.

Stenodesmus
    tenuicuspis (Mitt.) Jaeg.

Stenodictyon
    bisodalense Allen, Crosby & Magill
    wrightii (Sull. & Lesq.) Crosby (S. nitidum (Mitt.) Jaeg.)

Thamniopsis
    cruegeriana (C. Müll.) Buck (Hookeriopsis)
    pendula (Hook.) Fleisch.
    sinuata (Mitt.) Buck (Hookeriopsis)
    undata (Hedw.) Buck (Hookeriopsis, H. crispa (C. Müll.) Jaeg.)

Trachyxiphium
    aduncum (Mitt.) Buck (Hookeriopsis)
    guadalupensis (Brid.) Buck (Hookeriopsis, H. falcata (Hook.) Jaeg., H.
        gracilis (Mitt.) Jaeg.)
    subfalcatum (Hampe) Buck (Hookeriopsis, H. scabrella (Mitt.) Jaeg.)

vagum (Mitt.) Buck (*Hookeriopsis*)

Calymperaceae

[References: Reese (1961, 1977, 1978, 1979, 1983, 1987a, b, 1993).]

Calymperes
   afzelii Sw. (*C. donnellii* Aust.)
   erosum C. Müll. (*C. sprucei* Besch.)
   lonchophyllum Schwaegr.
   palisotii Schwaegr. (*C. richardii* C. Müll.)

Syrrhopodon
   circinatus (Brid.) Mitt. (*S. subrigidus* Broth.)
   cryptocarpos Dozy & Molk.
   gaudichaudii Mont.
   graminicola Williams
   hornschuchii Mart.
   incompletus Schwaegr. var. *incompletus*
   leprieurii Mont. (*S. iridans* Mitt., *S. pallidus* Mitt., *S. sylvaticus* Mitt.)
   ligulatus Mont.
   lycopodioides (Brid.) C. Müll.
   parasiticus (Brid.) Besch.
   prolifer Schwaegr.
      var. *prolifer*
      var. *acanthoneuros* (C. Müll.) C. Müll. (*S. allionii* Broth., *S. macrophyllus* Broth., *S. subscaber* Broth.)
      var. *papillosus* (C. Müll.) Reese
      var. *tenuifolius* (Sull.) Reese
   rigidus Hook. & Grev.
   tortilis Hampe (*S. ulei* C. Müll.)

Catagoniaceae

[Reference: Lin (1984). Note: Buck and Ireland (1985) have placed *Catagonium* in its own family; previously the genus was associated with either the Phyllogoniaceae or Plagiotheciaceae.]

Catagonium
   brevicaudatum C. Müll. ex Broth. (*Eucatagonium politum* (Hook. f. & Wils.) Broth.)

Cryphaeaceae

[References: Manuel (1977c, 1981).]

Cryphaea

apiculata B.S.G. (*C. latifolia* Mitt.)
attenuata B.S.G.
brevipila Mitt.
fasciculosa Mitt.
jamesonii Tayl.
patens Hornsch. ex Mitt.
pilifera Mitt.
ramosa Wils.

Schoenobryum
  caripense (C. Müll.) Manuel (*Acrocryphaea*)
  julaceum Dozy & Molk. (*Acrocryphaea*)
  rubicaule (Mitt.) Manuel (*Acrocryphaea*)

Daltoniaceae

[Reference: Bartram (1931). Note: The genera included here have traditionally been placed in the Hookeriaceae *s.l.*; see comments under Callicostaceae.]

Calyptrochaeta
  mniadelphus (Spruce ex Mitt.) Churchill *comb. nov. Eriopus*
    mniadelphus Spruce ex Mitten, J. Linn. Soc., Bot. 12: 392. 1869.

Daltonia
  bilimbata Hampe
  gracilis Mitt.
  jamesonii Tayl.
  lindigiana Hampe
  longifolia Tayl.
  macrotheca Mitt.
  ovalis Tayl.
  pulvinata Mitt.
  stenophylla Mitt.
  tenuifolia Mitt.
  trachyodonta Mitt.

Leskeodon
  andicola (Mitt.) Broth.
  palmarum (Mitt.) Broth.

Dicranaceae

[References: Frahm (1978, 1981, 1983, 1984, 1987a, 1989, 1991); Giese and Frahm (1985); Griffin (1986a, 1989b); Hegewall (1978); Padberg and Frahm (1985).]

Anisothecium
  campylophyllum (Tayl.) Mitt.
  convolutum (Hampe) Mitt. (*Aongstroemia*)

hookeri (C. Müll.) Broth. (*A. jamesonii* Mitt.)
planinervium (Tayl.) Mitt.

Aongstroemia
filiformis (P. Beauv.) Wijk & Marg. (*A. Jamaicensis* C. Müll.)
julacea (Hook.) Mitt. (*A. vulcanica* (Brid.) C. Müll.

Atractylocarpus
longisetus (Hook.) Bartr.

Bryohumbertia
filifolia (Hornsch.) J.-P. Frahm (*Campylopus; C. porphyreo-dictyon*
(C. Müll.) Mitt.)

Campylopus
albidovirens Herz.
anderssonii (C. Müll.) Jaeg. (*C. insularis* Bartr.)
arctocarpus (Hornsch.) Mitt.
areodictyon (C. Müll.) Mitt. (*C. subconcolor* (Hampe) Mitt.)
argyrocaulon (C. Müll.) Broth. (*C. leucognodes* (C. Müll.) Par.)
asperifolius Mitt. (*C. trichophorus* Hampe ex Herz.; *Dicranodontium*)
capitulatus Bartr.
cavifolius Mitt.
concolor (Hook.) Brid.
cuspidatus (Hornsch.) Mitt. var. *dicnemioides* (C. Müll.) J.-P. Frahm
densicoma (C. Müll.) Par. var. *densicoma*
edithae Broth. (*C. harpophyllus* Herz.)
flexuosus (Hedw.) Brid. (*C. brachyphyllus* Mitt.)
fragilis (Brid.) B.S.G. var. *fragilis* (*C. fimbriatus* Mitt.)
heterostachys (Hampe) Jaeg. (*C. annotinus* Mitt.)
huallagensis Broth. var. *huallagensis*
jamesonii (Hook.) Jaeg.
lamellinervis (C. Müll.) Mitt. var. *lamellinervis*
nivalis (Brid.) Brid. var. *nivalis* (*C. chrismarii* (C. Müll.) Mitt., *C.*
krauseanus (Hampe ex Lor.) Par., *C. suboblongus* Herz. var.
multicapsularis (C. Müll.) J. P.-Frahm)
pauper (Hampe) Mitt. (*C. rosulatus* (Hampe) Mitt.)
pilifer Brid.
    ssp. *pilifer* var. *pilifer*
    ssp. *pilifer* var. *lamellatus* (Mont.) Gradstein & Sipman (*C.*
      amellatus Mont.)
    ssp. *galapagensis* (J.-P. Frahm & Sipman) J.-P. Frahm (*C.*
      galapagensis J.-P. Frahm & Sipman)
pittieri Williams
reflexisetus (C. Müll.) Broth. (*C. ptychotheca* Herz.)
richardii Brid. (*Thysanomitrium*)
sharpii J.-P. Frahm, Horton & Vitt
subjugorum Broth.

tallulensis Sull. & Lesq.
trichophylloides Thér. in Herz.
trivialis C. Müll. ex Britt.
zygodonticarpus (C. Müll.) Par.

Excluded:
clavatus (R. Brown) Wils. (*C. leptodus* Mont.; *Thysanomitrium
leptopodum* (Mont.) Dix. of Steere, 1948)
introflexus (Hedw.) Brid.
microphyllinus Broth.
surinamensis C. Müll. (*C. subleucogaster* (C. Müll.) Jaeg.)
terebrifolius (C. Müll.) Jaeg.

Chorisodontium
mittenii (C. Müll.) Broth.
wallisii (C. Müll.) Broth.
var. *wallisii*
var. *speciosum* (Hook. f. & Wils.) Frahm

Dicranella
angustifolium Mitt.
elata Schimp ex Mitt.
harrisii (C. Müll.) Broth.
hilariana (Mont.) Mitt.
luteola Mitt.
osculatiana De Not.
perrottetii (Mont.) Mitt.
sericea Bartr.
vaginata (Hampe) Jaeg. (*Anisothecium*)

Dicranoweisia
fastigiata (Mitt.) Par.

Dicranum
frigidum C. Müll.

Holodontium
inerme (Mitt.) Broth.

Holomitrium
arboreum Mitt.
crispulum Mart.
flexuosum Mitt.
pulchellum Mitt.
standleyi Bartr.
tortuosum Mitt.
undulatum Mitt.

Leucoloma
  macrodon (Hook.) Jaeg.
  mollissimum Mitt.
  serrulatum Brid.

Microcampylopus
  curvisetus (Hampe) Giese & J.-P. Frahm (*Campylopodium pusillum*
  (Schimp.) R. S. Williams)

Microdus
  densus (Hook.) Besch.
  exiguus (Schwaegr.) Besch.
  longirostris (Schwaegr.) Schimp. (*Dicranella*)
  rubrisetus Bartr.

Oreoweisia
  erosa (Mitt.) C. Müll. (*O. ampliata* Mitt., *O. lechleri* (C. Müll.) Par. *O.*
  ligularis Mitt.)

Pilopogon
  guadeloupensis (Brid.) J.-P. Frahm (*P. gracilis* (Hook.) Brid.)
  laevis (Tayl.) Thér. (*Pilopogonella; P. nanus* Hampe, *P. piliferus*
      Hampe)
  longirostratus Mitt.
  macrocarpus Broth.
  peruvianus (Williams) J. -P. Frahm

Schliephackea
  prostrata C. Müll.

Symblepharis
  fragilis Mitt.
  lindigii Hampe
  vaginata (Hook.) Wijk & Marg. (*S. helicophylla* Mitt.)

Excluded: *Paraleucobryum albicans* (Schwaegr.) Loeske (*Leucobryum
megalophyllum* Mitt.; identity of this name is unknown).

Ditrichaceae

[Reference: Burley and Pritchard (1990).]

Ceratodon
  purpureus (Hedw.) Brid.
  stenocarpus B.S.G.

Distichium
  capillaceum (Hedw.) B.S.G.

Ditrichum
  crinale (Tayl.) O. Kuntze
  gracile (Mitt.) O. Kuntze
  rufescens (Hampe) Hampe
  strictum (Hook. f. & Wils.) Hampe

Encalyptaceae

[Reference: Horton (1983).]

Encalpyta
  asperifolia Mitt.
  ciliata Hedw. (*E. coarctata* Mitt.)

Entodontaceae

[Reference: Buck (1980a).]

Entodon
  beyrichii (Schwaegr.) C. Müll. (*E. erythropus* Mitt.)
  concinnus (De Not.) Par.
  hampeanus C. Müll.
  jamesonii (Tayl.) Mitt.
  pallescens (C. Müll.) Mitt.
  pallidisetus Mitt.
  wagneri Lor.

Erythrodontium
  longisetum (Hook.) Par.
  squarrosum (Hampe) Par. (*E. consanguineum* (Hampe) Par.)

Mesonodon
  flavescens (Hook.) Buck (*Campylodontium onustum* (Hampe) Jaeg.)

Erpodiaceae

[Reference: Crum (1972).]

Erpodium
  coronatum (Hook. f. & Wils.) Mitt.
  domingense (Brid.) C. Müll.

Eustichiaceae

[Reference: Crum (1984a).]

Diplostichum

longirostre (Brid.) Brid. (*Eustichia jamesonii* (Tayl.) C. Müll., *E. poeppigii* (C. Müll.) Par.)

Fabroniaceae

[References: Buck (1983); Buck and Crum (1978).]

Fabronia
  ciliaris (Brid.) Brid.
    var. *polycarpa* (Hook.) Buck (*F. polycarpa* Hook.)
    var. *wrightii* (Sull.) Buck (*F. andina* Mitt., *F. nivalis* Mont.; *Fabronidium espinosae* Herz.)
  jamesonii Tayl.

Fissidentaceae

[References: Bruggeman-Nannenga (1979); Pursell (1984; per. comm.); Pursell and Vital (1986); Pursell *et al.* (1988).]

Fissidens
  allionii Broth.
  altolimbatus Broth.
  asplenioides Hedw.
  crispus Mont.
  cylindraceus Mitt.
  diplodus Mitt. (*F. muriculatus* Spruce ex Mitt.)
  elegans Brid.
  flavinervis Mitt.
  guianensis Mont.
  hydropogon Spruce ex Mitt.
  inaequalis Mitt.
  intermedius C. Müll.
  intramarginatus (Hampe) Mitt.
  microcladus Thwait. & Mitt. (*F. garberi* Lesq. & Sull.)
  mollis Mitt. (*F. macrophyllus* Mitt.)
  polypodioides Hedw.
  prionodes Mont. sensu Florschütz, 1964.
  ramicola Broth.
  repandus Wils.
  reticulous (C. Müll.) Mitt.
  rigidulus Hook. f. & Wils.
  scalaris Mitt.
  scariosus Mitt.
  steerei Grout
  stenopteryx Besch.
  weirii Mitt.
    var. *weirii* (*F. howellii* Bartr.)
    var. *hemicraspedophyllus* (Card.) Pursell

zollingeri Mont. (*F. kegelianus* C. Müll.)

Funariaceae

[References: Fife (1987), Robinson and Delgadillo M. (1973).]

Entosthodon
    acaulis (Hampe) Fife (*E. steerei* Fife)
    andicola Mitt.
    bonplandii (Hook.) Mitt.
    jamesonii (Tayl.) Mitt. (*Funaria acidota* (Tayl.) C. Broth.)
    laevis (Mitt.) Fife (*Funaria suberecta* Mitt.)
    laxus (Wils. & Hook. f.) Mitt. (*Physcomitrium benoistii* Thér.)
    lindigii (Hampe) Mitt.
    mathewsii Hook. f. in Hook. (*Funaria eurystoma* (Mitt.) Broth.)
    obtusifolius Hook. f. in Hook. (*Funaria acutifolius* (Hampe ex C.
        Müll.) Broth., *E. longicollis* Mitt.)

Funaria
    calvescens Schwaegr.
    hygrometrica Hedw.

Neosharpiella
    turgida (Mitt.) H. Robins. & C. Delg. (*Physcomitrium*)

Grimmiaceae

[References: Bremer (1980a, b, 1981); Churchill (1981); Deguchi (1987). Earlier classifications have placed *Ptychomitrium* in its own family close to the Orthotrichaceae.]

Grimmia
    affinis Hornsch.
    allionii Broth. nom. nud.
    benoistii Thér.
    cinerea Thér.
    columbica De Not.
    consobrina Kunz.
    fusco-lutea Hook.
    longirostris Hook. (*G. peruviana* Sull.)
    ovalis (Hedw.) Lindb. (*G. ovata* Web. & Mohr)
    stenopyxis Thér.

Ptychomitrium
    chimborazense (Spruce) Jaeg.

Racomitrium

crispipilum (Tayl.) Jaeg. (*Racomitrium vulcanicum* Lor., *syn. nov.*, Moosstud. 163. 1864. *R. crispulum* (Hook. f. & Wils.) Hook. f. & Wils. of reports). var. *brevifolium* Thér. *lanuginosum* (Hedw.) Brid.

Schistidium
angustifolium (Mitt.) Herz. (*Grimmia*)
apocarpum (Hedw.) Bruch & Schimp. in B.S.G. (*Grimmia; S. andinum* (Mitt.) Herz., *G. saxatilis* Mitt.)
rivulare (Brid.) Podp. (*Grimmia rivulariopsis* R. S. Willams)

Hedwigiaceae

[Reference: De Luna and Buck (1991). Note: The genus *Rhacocarpus* has been placed in its own family.]

Braunia
cirrhifolia (Wils.) Jaeg.
nephelogenes De Luna & Buck
plicata (Mitt.) Jaeg.
secunda (Hook.) B.S.G.

Hedwigia
ciliata (Hedw.) P.-Beauv.

Hedwigidium
integrifolium (P.-Beauv.) Dix. in C. Jens. (*H. imberbe* (Sm.) B. S. G.)

Hookeriaceae

[Note: See comments under Callicostaceae.]

Hookeria
acutifolia Hook. in Grev.

Hydropogonaceae

[Reference: Churchill (1991a).]

Hydropogon
fontinaloides (Hook.) Brid.

Hylocomiaceae

[Note: *Pleurozium* has been placed in various families in the past including the Amblystegiaceae and Entodontaceae; see Buck (1980a).]

Pleurozium
    schreberi (Brid.) Mitt. (*Calliergonella*)

Hypnaceae

[References: Ando (1978); Ando and Higuchi (1984); Buck (1984, 1987b); Ireland (1991); Nishimura (1985); Nishimura and Ando (1986); Nishimura *et al.* (1984).]

Chryso-hypnum
    diminutivum (Hampe) Buck (*M. perspicuum* (Hampe) Card., *M. subthelistegum* (Card.) Card., *M. thelistegum* (C. Müll.) Card.)

Ctenidium
    malacodes Mitt.

Ectropothecium
    aeruginosum (C. Müll.) Mitt.
    leptochaeton (Schwaegr.) Buck (*E. apiculatum* Mitt., *E. globitheca* (C. Müll.) Mitt.)

Hypnum
    amabile (Mitt.) Hampe
    cupressiforme Hedw. var. *lacunosum* Brid.
    polypterum (Mitt.) Broth. (*Caribaeohypnum*)

Isopterygium
    sprucei (Mitt.) Buck (*Syringothecium*)
    tenerifolium Mitt. (*I. longisetum* Broth.)
    tenerum (Sw.) Mitt.

Mittenothamnium
    andicolum (Hook.) Card.
    incompletum (Fleisch.) Card.
    jamesonii (Tayl.) Card.
    langsdorffii (Hook.) Card.
    oxyrrhynchioides (Broth.) Steere
    reptans (Hedw.) Card. (*M. pallidum* (Hook.) Card.)
    rivulare (Broth.) Steere
    subobscurum (Hampe) Card.
    substriatum (Hampe) Card.
    trichocladon (Ren. & Card.) Card.
    volvatum (Hampe) Card.

Phyllodon
    truncatulus (C. Müll.) Buck (*Glossadelphus*)

Platygyriella
    densa (Hook.) Buck (*Bryo-sedgwickia; Erythrodontium*)

Pylaisiella
  falcata (B.S.G.) Ando (*Stereodon hamatus* Mitt.)

Racopilopsis
  trinitensis (C. Müll.) Britt. & Dix.

Taxiphyllum
  laevifolium (Mitt.) Broth. (*Glossadelphus*)
  taxirameum (Mitt.) Fleisch. (*T. planissimum* (Mitt.) Broth.)

Vesicularia
  vesicularis (Schwaegr.) Broth. (*V. amphibola* (Mitt.) Broth.)
  suburceolata (Hampe & Lor.) Broth.

Hypopterygiaceae

Hypopterygium
  lehmannii Besch.
  tamariscinum (Hedw.) Brid.

Lembophyllaceae

[Note: *Porotrichodendron* has been traditionally been placed in the Lembophyllaceae, but appears better placed in the Thamnobryaceae.]

Leptodontaceae

[Reference: Buck (1980b). Note: *Pseudocryphaea* has traditionally been placed in the Cryphaeaceae.]

Pseudocryphaea
  domingensis (Spreng.) Buck (*P. flagellifera* (Brid.) Britt.)

Lepyrodontaceae

Lepyrodon
  tomentosus (Hook.) Mitt. (*L. suborthostichus* (C. Müll.) Hampe)

Leskeaceae

[Reference: Buck and Crum, 1990.]

Leskea
  angustata Tayl.
  plumaria Mitt. (*Rauia, Rauiella*)
  teretiuscula Mitt. (*Rauia, Rauiella*)

180 S. P. Churchill

Leucobryaceae

[References: Salazar Allen (1991, 1993).]

Leucobryum
  albidum (P. Beauv.) Lindb.
  antillarum Schimp. ex Besch.
  crispum C. Müll.
  giganteum C. Müll. (*L. longifolium* Hampe)
  martianum (Hornsch.) Hampe

Leucophanes
  molleri C. Müll. (*L. calymperatum* C. Müll., *L. mittenii* Card.)

Octoblepharum
  albidum Hedw.
  cocuiense Mitt.
  pulvinatum (Dozy & Molk.) Mitt.

Leucomiaceae

[Reference: Allen (1987). Note: *Rhynchostegiopsis* has been previously associated with the Hookeriaceae *s. l.*]

Leucomium
  strumosum (Hornsch.) Mitt. (*L. compressum* Mitt., *L. lignicola* Spruce
    & Mitt.)

Rhynchostegiopsis
  tunguraguana (Mitt.) Broth.

Meteoriaceae

[References: Allen and Crosby (1986b); Griffin (1986b); Lewis (1992); Manuel (1977a, b); Menzel (1991); Robinson (1967); Visnadi and Allen (1991).]

Barbella
  trichophora (Mont.) Fleisch. ex Broth. (*B. cubensis* (Sull.) Broth.;
    Squamidium)

Lindigia
  debilis (Wils. ex. Mitt.) Jaeg. (*Eriodon brevisetus* Bartr.)

Meteoridium
  remotifolium (C. Müll.) Manuel (*Meteoriopsis, M. consimilis*
    (Hampe) Broth., *M. breviseta* (Mitt.) Broth.)
  tenuissima (Hook. f. & Wils.) Lewis (*Barbella*)
Meteorium

sinuatum (C. Müll.) Mitt. *(M. illecebrum* Sull.)
teres Mitt.

Papillaria
  deppei (Hornsch.) Jaeg.
  mponderosa (Tayl.) Broth.
  nigrescens (Hedw.) Jaeg.
  penicillata (Dozy & Molk.) Broth. (*P. laevifolia* (Mitt.) Broth.)

Pilotrichella
  allionii Broth.
  flexilis (Hedw.) Ångstr. (*P. turgescens* (C. Müll.) Besch.)
  hexasticha (Schwaegr.) Jaeg.
  pentasticha (Brid.) Wijk & Marg. (*P. versicolor* (C. Müll.) Jaeg., *P. viridis* (C. Müll.) Jaeg.)
  quitensis (Mitt.) Jaeg.

Squamidium
  isocladum (Ren. & Card.) Broth.
  leucotrichum (Tayl.) Broth. (*S. caroli* (C. Müll.) Broth.)
  livens (Schwaegr.) Broth. (*S. filiferum* (C. Müll.) Broth.)
  macrocarpum (Spruce ex Mitt.) Broth.
  nigricans (Hook.) Broth. (var. *compactum* Thér., var. *gracile* Broth.)

Zelometeorium
  allionii Manuel
  patens (Hook.) Manuel (*Meteoriopsis tovariensis* (C. Müll.) Broth.)
  patulum (Hedw.) Manuel (*Meteoriopsis; M. anderssonii* (C. Müll.) Broth.)
  recurvifolium (Hornsch. in Mart.) Manuel (*Meteoriopsis; M. onusta* (Spruce) Broth.)

Mniaceae

[Reference: Koponen (1979).]

Plagiomnium
  rhynchophorum (Hook.) T. Kop. (*Mnium longirostre* Brid., var. *longirostre* Brid. & var. *ligulatum* (C. Müll.) Steere)

Myriniaceae

[Reference: Buck and Crum (1978).]

Helicodontium
  laevisetum jaeg.
  minutum (Mitt.) Jaeg.

oblique-rostratum (Mitt.) Jaeg.

Neckeraceae

[Reference: Sastre-De Jesús (1987). Note: The genera *Pinnatella, Porotrichum* (including *Porothamnium*), and *Thamnobryum* have been placed in the Thamnobryaceae (Sastre-De Jesús, op. cit.).]

Isodrepanium
  lentulum (Wils.) Britt.

Neckera
  andina Mitt.
  benoistii Thér.
  chilensis Schimp. ex Mont. (*N. jamesonii* Tayl., *N. osculatiana* De Not.)
  ehrenbergii C. Müll. (*spruceana* Mitt. as *Neckeradelphus*)
  obtusifolius Tayl. (*Neckeradelphus*)
  scabridens C. Müll. (*N. lindigii* Hampe)
  urnigera C. Müll. (*N. mollusca* Mitt.)

Neckeropsis
  disticha (Hedw.) Kindb.
  undulata (Hedw.) Reichdt.

Orthotrichaceae

[References: Grout (1946); Lewinsky (1976, 1984, 1987); Malta (1926) Robinson (1975); Vitt (1979, 1980).]

Amphidium
  tortuosum (Hornsch.) H. Robins. (*A. cyathicarpum* (Mont.) Broth.)

Groutiella
  apiculata (Hook.) Crum & Steere (*Craspedophyllum; G. mucronifolia* (Hook. & Grev.) Crum & Steere)
  chimborazense (Spruce ex Mitt.) Florsch.
  fragile (Mitt.) Steere (*Craspedophyllum*)
  tomentosa (Hornsch.) Wijk & Marg.
  tumidula (Mitt.) Vitt
  wagneriana (C. Müll.) Crum & Steere (*Craspedophyllum*)

Macrocoma
  tenue (Hook. & Grev.) Vitt. subsp. *sullivantii* (C. Müll.) Vitt (*M. hymenostomum* (Mont.) Grout, of reports)

Macromitrium
  aureum C. Müll.
  cirrosum (Hedw.) Brid.

var. *cirrosum* (*M. microtheca* Mitt.)
var. *stenophyllum* (Mitt.) Grout (*M. laevisetum* Mitt.)
constrictum Hampe & Lor.
contextum Hampe
crispatulum Mitt.
cylindricum Mitt.
divaricatum Mitt.
drewii H. Robins.
flexuosum Mitt.
frondosum Mitt.
guatemalense C. Müll. (*M. penicillatum* Mitt.)
homalacron C. Müll.
huigrense Williams
longifolium (Hook.) Brid.
mittenianum Steere
oblongum (Tayl.) Mitt.
osculatianum De Not.
ovale Mitt.
pentastichum C. Müll.
perreflexum Steere
podocarpi C. Müll.
portoricense Williams
punctatum (Hook. & Grev.) Brid.
richardii Schwaegr. (*M. didymodon* Schwaegr., *M. rhabdocarpum* Mitt.)
rimbachii Herz.
scoparium Mitt. (*M. trichophyllum* Mitt.)
serrulatum Mitt.
squarrosum C. Müll. hom. illeg.
stellulatum (Hornsch.) Brid.
subcrenulatum Broth. in Herz.
sublaeve Mitt.
subscabrum Mitt.
trachypodium Mitt.
ulophyllum Mitt.

Orthotrichum
aequatoreum Mitt.
diaphanum Schrad. ex Brid.
elongatum Tayl.
latimarginatum Lewinsky
laxifolium Wils. in Mitt.
mandonii Schimp. ex Hampe
pariatum Mitt.
pungens Mitt.
pycnophyllum Schimp. in C. Müll. var. *pycnophyllum* (*O. apiculatum* Mitt., *O. epibryum* De Not., *O. rubescens* Mitt., *O. wagneri* Lor.)
var. *verrucosum* (C. Müll.) Lewinsky

rupestre Schleich. ex Schwaegr. (*O. nivale* Spruce; *O. striatum* Hedw.
   of reports)
steerei Lewinsky (*O. undulatum* Mitt., hom. illeg.)
subulatum Mitt.
trachymitrium Mitt. (*O. patulum* Mitt.)

Schlotheimia
   angustata Mitt.
   krausei Hampe & Lor.
   rugifolia (Hook.) Schwaegr. (*S. jamesonii* (Arnott) Brid.)
   torquata (Hedw.) Brid.

Zygodon
   altarensis Broth. in Malta
   fasciculatus Mitt.
   fragilis H. Robins.
   liebmannii Schimp. (*Z. brevicollis* Mitt.)
   nivalis Hampe
   peruvianus Sull. (*Z. goudotii* Hampe)
   pichinchensis (Tayl.) Mitt.
   quitensis Mitt. var. *integrifolius* Malta
   reinwardtii (Hornsch.) Braun in B.S.G. (*Z. subdenticulatus* Hampe)
   rufescens (Hampe) Broth. in Par.
   squarrosus (Tayl.) C. Müll.
   stenocarpus Tayl.
      var. *stenocarpus*
      var. *linearifolius* (Mitt.) Malta
   subsquarrosus Broth. in Malta

Phyllodrepaniaceae

Mniomalia
   viridis (Mitt.) C. Müll.

Phyllogoniaceae

[Reference: Lin (1983).]

Phyllogonium
   fulgens (Hedw.) Brid.
   viscosum (P. Beauv.) Mitt.

Plagiotheciaceae

[References: Buck and Ireland (1985, 1989).]

Plagiothecium

conostegium Herz.
lucidum (Hook. f. & Wils.) Par.
novo-granatense (Hampe) Mitt.

Polytrichaceae

[References: Hyvönen (1989), Menzel (1985, 1986); Nyholm (1971); Robinson (1967); Smith (1971, 1975).]

Atrichum
polycarpum (C. Müll.) Mitt. (*A. pastasanum* Mitt., *A. planifolium* (C. Müll.) Jaeg.)

Notoligotrichum
trichodon (Hook. f. & Wils.) G. L. Sm. (*Psilopilum*)

Pogonatum
campylocarpum (C. Müll.) Mitt. (*P. andinum* (Hampe) Mitt., *P. obscuratum* Mitt., *P. varians* (Hampe) Mitt.)
neglectum (Hampe) Jaeg.
perichaetiale (Mont.) Jaeg. ssp. *oligodus* (C. Müll.) Hyvönen (*P. plurisetum* (C. Müll.) Broth., *P. polycarpum* (C. Müll.) Par.)
semipellucidum (Hampe) Mitt.
tortile (Sw.) Brid.

Polytrichadelphus
aristatus (Hampe) Mitt.
bolivianus Herz.
giganteus (Hook.) Mitt.
longisetus (Hook.) Mitt.
purpureus Mitt.
rubiginosus Mitt.
subrubescens (Thér.) G. L. Sm. (*Polytrichum*)
valenciae (C. Müll.) Par.

Polytrichastrum
tenellum (C. Müll.) G. L. Sm.

Polytrichum
juniperinum Hedw. (*P. aequinoctiale* (Lor.) Broth., *P. antillarum* Rich. in Brid., *P. aristiflorum* Mitt., ? *P. chimborassi* Lor., *P. conforme* Mitt., *P. cuspidigerum* Schimp. ex Britt., *P. substrictum* Hampe in Par.)
rubescens Mitt.

Pottiaceae

[References: Churchill (1990); Delgadillo (1975); Long (1979), Mishler (in press, pers. comm.), Redfearn (1991), Steere (1986); Zander (1972, 1977, 1978a, b, 1979, 1981a, b, 1983, 1986, 1989).]

Aloina
  rigida (Hedw.) Limpr. (*A. calceolifolia* (Mitt.) Broth.)

Aloinella
  cucullifera (C. Müll.) Steere

Anoectangium
  peruvianum Sull.

Barbula
  arcuata Griff. (*B. subulifolia* Sull.)
  brachymenia (Mitt.) Jaeg.
  calyculosa (Mitt.) Jaeg.
  costata (Mitt.) Jaeg.
  decolorans Hampe
  ecuadorensis Broth.
  granulosa Thér.
  hyalinobasis Broth.
  indica (Hook.) Spreng. (*B. cruegeri* Sond. ex C. Müll.)
  ligularis (Mitt.) Jaeg.
  linguaecuspis Broth.
  taylorii Bartr. & Steere hom. illeg.
  torquata Tayl.
  vulcanica Lor.

Bryoerythrophyllum
  campylocarpum (C. Müll.) Crum (*Barbula; Didymodon arcuatus* (Mitt.) Broth.)
  inaequalifolium (Tayl.) Zander (*Barbula*)
  jamesonii (*Leptodontium subplanifolium* Thér.)

Calyptopogon
  mnioides (Schwaegr.) Broth.

Desmatodon
  bellii Bartr.

Didymodon
  australasiae (Hook. & Grev.) Zander (*Barbula; Trichostomopsis*)
  chimborazensis (Mitt.) Broth.
  Tortula humida Mitt. to be transfered by Zander
  inundatus (Mitt.) Broth.
  laevigatus (Mitt.) Zander (*Barbula*)

montanus (Mitt.) Broth. *Barbula pruinosa* (Mitt.) Jaeg. to be transfered
   by Zander
revolutus (Card.) Will. (*Husnotiella*) *Barbula campylocarpa* (Tayl.) C.
   Müll. to be transfered under another name by Zander
tophaceus (Brid.) Lisa
vinealis (Brid.) Zander (*Barbula; B. rectifolia* Tayl.)

Dolotortula
   mniifolia (Sull.) Zander (*Tortula*)

Gymnostomiella
   vernicosa (Hook.) Fleisch. (*G. orcuttii* Bartr.)

Gymnostomum
   aeruginosum Sm. (*G. calcareum* Nees, Hornsch. & Sturm)
   aestivum Hedw. (*Anoectangium, A. calidum* Mitt., *A. compactum*
      Schwaegr., *A. euchloron* (Schwaegr.) Mitt.; *Zygodon tenellus* Mitt.)

Gyroweisia
   benoisti Thér.

Hydrogonium
   taylori W. A. Weber

Hymenostomum
   breutelii (C. Müll.) Broth.
   densirete Thér.
   subacaule (Mitt.) Par.

Hymenostylium
   recurvirostrum (Hedw.) Dix. (*Gymnostomum, H. stillicidiorum* (Mitt.)
      Broth.)

Hyophila
   grossidens Broth.
   involuta (Hook.) Jaeg. (*H. tortula* (Schwaegr.) Hampe)
   microcarpa (Schimp. ex Besch.) Broth.

Leptodontium
   acutifolium Mitt.
   anomalum Dix. & Thér.
   brachyphyllum Broth. & Thér.
   capituligerum C. Müll. (*L. calymperoides* Thér.)
   flexifolium (Dicks. ex With.) Hampe in Lindb. (*L. filiformis* (Lor.)
      Steere)
   longicaule Mitt. var. *longicaule*
   luteum (Tayl.) Mitt. var. *microruncinatum* (Dus.) Zander
   molle C. Müll.

pungens (Mitt.) Kindb. (*L. acutifolium* Bartr.)
stoloniferum Zander
syntrichioides (C. Müll.) Kindb.
viticulosoides (P. Beauv.) Wijk & Marg.
  var. *viticulosoides* (*L. densifolium* (Mitt.) Mitt.)
  var. *flagellaceum* (Bartr.) Zander
  var. *sulphureum* (Lor.) Zander (var. *panamense* (Lor.) Zander;
    L. cirrhifolium Mitt.)
wallisii (C. Müll.) Kindb.

Molendoa
  andina (Mitt.) Broth.
  sordida (Mitt.) Steere
  subobtusifolia Broth. nom. nud.

Morinia
  ehrenbergiana (C. Müll.) Thér.
    var. *ehrenbergiana*
    var. *elongata* (Wils. in Mitt.) Zander (*Barbula elongata* Wils.
      in Mitt.; *M. ecuadorensis* Bartr.)

Oxystegus
  tenuirostris (Hook. & Tayl.) A. Sm. (*Trichostomum*)

Phascum
  cuspidatum Hedw.

Pleurochaete
  ecuadoriensis Broth.
  squarrosa (Brid.) Lindb. (*P. luteola* (Besch.) Thér.)

Pseudocrossidium
  excavatum (Mitt.) Williams
  replicatum (Tayl.) Zander (*Barbula*)
  steerei Churchill

Rhamphidium
  brevifolium (Hampe & Lor.) Broth.
  dicranoides (C. Müll.) Par.

Sagenotortula
  quitoensis (Tayl. in Hook.) Zander (*Tortula*)

Scopelophila
  ligulata (Spruce) Spruce (*Merceya; M. agoyanensis* (Mitt.) C. Müll.,
    M. cataractae (Mitt.) C. Müll.)

Streptopogon

calymperes Geh. (*S. rigidus* Mitt. ex Broth.)
cavifolius Mitt.
clavipes Spruce ex Mitt.
erythrodontus (Tayl.) Wils.
    var. *erythrodontus*
    var. *intermedius* Salmon

Tortella
contortifolia (Mitt.) Broth. in Par.
humilis (Hedw.) Jenn.
simplex H. Robins.

Tortula
aculeata (Wils.) Mitt.
amphidiacea (C. Müll.) Broth. (*T. caroliniana* Andr.)
bogotensis (Hampe) Mitt. (*T. decidua* Mitt.)
crenata Mitt.
denticulata (Tayl.) Mitt.
elongata (Wils.) Mitt. hom. illeg. (*Barbula*)
fragilis Tayl.
glacialis (C. Müll.) Mont. in Gay
leiostoma Herz. (*T. limbata* (Mitt.) Mitt., hom. illeg.)
napoana De Not.
papillosa Wils. in Spruce
pichinchensis Tayl.
princepes De Not.

Trachyodontium
zanderi Steere

Trichostomum
aequatoriale (Spruce) Dix.
bellii Bartr.
brachydontium Bruch
jamaicense (Mitt.) Jaeg.
linealifolium C. Müll.
perviride Broth.

Triquetrella
spiculosa Thér.

Weissia
controversa Hedw.
umbrosa Mitt.

Prionodontaceae

[Reference: Griffin (1970, in Sharp et al., in press).]

Prionodon
  densus Hedw. (*P. lycopodium* (C. Müll.) Jaeg. var. *abbreviatum*
    Thér., *P. undulatus* Mitt.)
  fusco-lutescens Hampe (*P. divaricatus* Mitt.)
  jamesonii C. Müll.
  luteo-virens (Tayl.) Mitt. (*P. laeviusculus* Mitt., *P. longissimus* Ren. &
    Card., *P. lycopodium* (C. Müll.) Jaeg., *P. patensissimus* Besch.,
    P. rigidus Ren. & Card.)
  lycopodioides Hampe

Pterobryaceae

[References: Arzeni (1954); Buck (1991); Churchill (1988a). Note: Buck (1989) replaced *Leucodontopsis* with an earlier name - *Henicodium*.]

Calyptothecium
  duplicatum (Schwaegr.) Broth.
  humile (Mitt.) Broth.

Henicodium
  geniculata (Mitt.) Buck (*Leucodontopsis*)

Hildebrandtiella
  guyanensis (Mont.) Buck (*Orthostichidium excavatum* (Mitt.) Broth.,
    O. pentagonum (Hampe & Lor.) C. Müll.

Orthostichopsis
  auricosta (C. Müll.) Broth.
  incertus Thér.
  praetermissa Buck (*O. crinita* (Sull.) Broth., of reports)
  tetragona (Hedw.) Broth.

Pireella
  cymbifolia (Sull.) Card.
  pergemmescens H. Robins.
  pohlii (Schwaegr.) Card.
  trichomanoides (Mitt.) Card.

Pterobryon
  densum (Schwaegr.) Hornsch.

Renauldia
  obovata Thér.

Racopilaceae

Racopilum

intermedium Hampe
tomentosum (Hedw.) Brid.

Rhacocarpaceae

Rhacocarpus
 excisus (C. Müll.) Par.
 purpurascens (Brid.) Par. (*R. humboldtii* (Hook.) Lindb.)

Rhegmatodontaceae

[Reference: Eakin (1976).]

Rhegmatodon
 polycarpus (Griff.) Mitt. (*R. schlotheimioides* Spruce in Mitt.)

Rhizogoniaceae

[References: Churchill and Buck (1982); Churchill and Matteri (in prep.).]

Leptotheca
 boliviana Herz.

Pyrrhobryum
 mnioides (Hook.) Manuel (*Rhizogonium*)
 spiniforme (Hedw.) Mitt. (*Rhizogonium*)

Rhizogonium
 lindigii Hampe

Rigodiaceae

[References: Crum (1984a); Zomlefer (1993); Zomlefer and Buck (1990).]

Rigodium
 toxarion (Schwaegr.) Jaeg. (*R. solutum* (Tayl.) Par.)

Seligeriaceae

[Reference: Bartlett and Vitt (1986).]

Blindia
 magellanica Schimp. in C. Müll. (*B. curviseta* Mitt.)

Sematophyllaceae

[Reference: Ochyra (1982); Tixier (1977).]

Acroporium
guianense (Mitt.) Broth.
pungens (Hedw.) Broth.
  var. *pungens*
  var. *stillicidiorum* (Broth.) Steere
ulicinum (Mitt.) Crum (*Schraderobryum*)

Allioniellopsis
cryphaeoides (Broth.) Ochyra (*Allioniella*)

Aptychella
proligera (Broth.) Herz.

Meiothecium
andinum Mitt.
boryanum (Mont.) Mitt.
negrense Spruce in Mitt.

Pterogonidium
liliputanum Broth.
pulchellum (Hook.) C. Müll. in Broth.

Sematophyllum
adnatum (Michx.) E. G. Britt.
andinum Mitt. (*Rhaphidorrhynchium*)
cuspidiferum Mitt.
decumbens Mitt. (*Rhaphidorrhynchium*)
esmeraldicum (C. Müll.) Broth.
napoanum (De Not.) Steere (*S. lindigii* (Hampe) Mitt.)
obliquerostratum Mitt. (*Rhaphidorrhynchium*)
subpinnatum (Brid.) Britt. (*S. caespitosum* Hedw. of authors;
    Rhaphidorrhynchium hedwigii (Mitt.) Broth.)
subsimplex (Hedw.) Mitt. (*Rhaphidorrhynchium*)

Taxithelium
planum (Brid.) Mitt.

Trichosteleum
ambiguum (Schwaegr.) Par.
fluviale (Mitt.) Jaeg.
papillosum (Hornsch.) Jaeg.
rubrisetum (Mitt.) Jaeg.

Sorapillaceae

[Reference: Allen (1981).]

Sorapilla
  sprucei Mitt.

Sphagnaceae

[References: Crum (1980, 1984b, 1987a, b, 1989, 1990); Griffin (1981).]

Sphagnum
  azuayense Crum
  barclayae Crum
  capillifolium (Ehrh.) Hedw. var. *tenerum* (Sull. & Lesq. ex Sull.)m Crum
  cuculliformes Crum
  cuspidatum Ehrh.
  laegaardii Crum
  magellanicum Brid. (*S. stewartii* Warnst.)
  meridense (Hampe) C. Müll.
  perichaetiale Hampe
  pulchricoma C. Müll.
  pylaesii Brid.
  sancto-josephense Crum & Crosby
  sparsum Hampe
  subsecundum Nees var. *rufescens* (Nees & Hornsch.) Hueb.

Splachnaceae

[Reference: Koponen (1977).]

Brachymitrion
  jamesonii Tayl. (*Tayloria*)
  laciniata (Spruce) A. Kop. (*Tayloria*)
  moritzianum (C. Müll.) A. Kop. (*Tayloria*)

Tayloria
  papulata C. Müll.
  scabriseta (Hook.) Mitt.

Splachnobryaceae

[Reference: Koponen (1981).]

Splachnobryum
  obtusum (Brid.) C. Müll.

Stereophyllaceae

[Reference: Buck and Ireland (1985).]

Entodontopsis
  leucostega (Brid.) Buck & Ireland (*Stereophyllum peruvianum* (Mont.) Mitt.)
  papillifera (Mitt.) Buck & Ireland (*Stereophyllum*)
  radicalis (Spruce ex Jaeg.) Buck (*Eriodon radicalis* Jaeg., *E. sprucei* Steere)

Pilosium
  chlorophyllum (Hornsch.) C. Müll. in Broth.

Sciuroleskea
  mittenii (Spruce ex Mitt.) Fleisch. ex Broth. (*Hypnum mittenii* Spruce ex Mitt.; *Moneurium mittenii* (Spruce ex Mitt.) C. Müll. ex Fleisch.; *Rozea roseorum* Williams)
  xanthophylla (Hampe & Lor.) Broth. (*Lescuraea xanthophylla* (Hampe & Lor.) Jaeg.)

Stereophyllum
  acutum Herz. hom. illeg.
  radiculosum (Hook.) Mitt.

Symphyodontaceae

[Reference: Buck & Ireland (1992); Steere (1982).]

Symphyodon
  imbricatifolius (Mitt.) S. P. Churchill, comb. nov.
    Leidopilum imbricatifolium Mitt., J. Linn. Soc., Bot. 12: 372. 1869.
  machrisianus (Crum) Buck & Ireland (*S. americanus* Steere)

Thamnobryaceae

[Reference: Sastre-De Jesús (1987). Note: Sastre-De Jesús (1987) has placed *Porothamnium* within *Porotrichum*, however not all transfers or synonyms have been given. *Porotrichodendron* has traditionally been placed in the Lembophyllaceae.]

Pinnatella
  caesia (Mitt.) Broth.
  minuta (Mitt.) Broth.

Porothamnium
  floridum (Tayl.) Williams
  gracile Broth.

Porotrichodendron
  nitidum (Hampe) Broth.
  superbum (Tayl.) Broth. in Herz.

Porotrichum
 expansum (Tayl.) Mitt. (*Porothamnium*)
 explanatum Mitt. (*Porothamnium*)
 filiferum Mitt. (*Porothamnium*)
 gymnopodum (Tayl.) Mitt. (*Porothamnium*)
 korthalsianum (Dozy & Molk.) Mitt. (*P. penicillidens* Steere)
 lancifrons (Hampe) Sastre-De Jesús (*Porothamnium imbricatum* (Spruce) Fleisch.)
 lehmannii Besch. (*Porothamnium*)
 lindigii (Hampe) Mitt. (*Pireella cavifolia* (Card. & Herz.) Card., fide I. Sastre-De Jesús)
 longirostre (Hook.) Mitt.
 mutabile Hampe (*P. insularum* Mitt., *P. variable* Hampe)
 protensum Ren. & Card. (*Porothamnium*)
 striatum Mitt.
 substriatum (Hampe) Mitt. (*P. plicatulum* Mitt.)

Thamnobryum
 fasciculatum (Hedw.) Sastre-De Jesús (*Porothamnium*)

Thuidiaceae

[Reference: Buck and Crum (1990); Gier (1980). Note: Species of *Rauiella* are here placed back in *Lesekea* (Leskeaceae).]

Cyrtohypnum
 campanulatum (Mitt.) Churchill, *comb. nov. Thuidium campanulatum* Mitt., J. Linn. Soc., Bot. 12: 574. 1869.
 cylindraceum (Mitt.) Churchill, *comb. nov. Thuidium cylindraceum* Mitt., J. Linn. Soc., Bot. 12: 574. 1869.
 espinosae (Herz.) Churchill, *comb. nov. Thuidium espinosae* Herz., Beih. Bot. Centralbl. 61B: 589. 1942.
 intermedium (Mitt.) Churchill, *comb. nov. Thuidium intermedium* Mitt., J. Linn. Soc., Bot. 12: 573. 1869.
 involvens (Hedw.) Buck & Crum (*Thuidium*)
 leptocladum (Tayl.) Buck & Crum (*Thuidium; T. brachythecium* (Hampe & Lor.) Jaeg.)
 minutulum (Hedw.) Buck & Crum (*Thuidium pauperum* (C. Müll.) Mitt.)

Heterocladium
 dimorphum (Brid.) B.S.G. (*H. squarrosulum* (Voit) Lindb.)
 plumulosum Herz.

Thuidium
 brasiliense Mitt.
 carantae (C. Müll.) Jaeg.
 delicatulum (Hedw.) Mitt.

peruvianum Mitt.
pseudo-delicatulum (C. Müll.) Jaeg.
pseudo-protensum (C. Müll.) Mitt.
tomentosum Besch. (*T. antillarum* Besch.)

Trachypodaceae

[Reference: Zanten (1959).]

Trachypus
    bicolor Reinw. & Hornsch. var. *viridulus* (Mitt.) Zant. (*T. viridulus* Mitt.)

9. LITERATURE CITED

Allen, B. H. 1981. A reevaluation of the Sorapillaceae. The Bryologist 84: 335–338.

Allen, B. H. 1986. The taxonomic status of *Hypnella punctata*. The Bryologist 89: 224–226.

Allen, B. H. 1987. A revision of the genus *Leucomium* (Leucomiaceae). Memoirs of the New York Botanical Garden 45: 661–677.

Allen, B. H. 1990. A revision of the genus *Crossomitrium* (Musci: Hookeriaceae). Tropical Bryology 2: 3–34.

Allen, B. H. and M. R. Crosby. 1986a. A revision of the genera *Pilotrichidium* and *Diploneuron* (Musci: Hookeriaceae). Journal of the Hattori Botanical Laboratory 61: 45–64.

Allen, B. H. and M. R. Crosby. 1986b. Revision of the genus *Squamidium* (Musci: Meteoriaceae). Journal of the Hattori Botanical Laboratory 61: 423–476.

Allen, B. H., M. R. Crosby and R. E. Magill. 1985. A review of the genus *Stenodictyon* (Musci). Lindbergia 11: 149–156.

Ando, H. 1978. *Pylaisiella falcata* (B.S.G.) Ando, a moss of East Asia-Latin American distribution. Phyta 1: 14–23.

Ando, H. and M. Higuchi. 1984. *Caribaeohypnum*, a new genus for a northern Latin American moss, *Hypnum polypterum* (Mitt.) Broth. Cryptogamie, Bryologie et Lichénologie 5: 7–14.

Arzeni, C. B. 1954. The Pterobryaceae of the southern United States, Mexico, Central America, and the West Indies. American Midland Naturalist 52: 1–67.

Balslev, H. 1988. Distribution patterns of Ecuadorean plant species. Taxon 37: 567–577.

Balslev, H. and T. de Vries. 1982. Diversidad de la vegetación en cuatro cuadrantes en el páramo arbustivo del Cotopaxi, Ecuador. Publicaciones del Museo Ecuatoriano de Ciencias Naturales 3: 20–32.

Balslev, H. and T. de Vries. 1991. Life forms and species richness in a bunch grass páramo on Mount Cotopaxi, Ecuador. Pp. 45–58. *In:* W. Erdelen, N. Ishwaran and P. Müller (eds.), Tropical Ecosystems. Margraf Scientific Books, Saarbrüken.

Balslev, H. and S. S. Renner. 1989. Ecuadorean forests east of the Andes. Pp. 287–296. *In:* L. B. Holm-Nielsen, I. C. Nielsen and H. Balslev (eds.), Tropical Forests: Botanical Dynamics, Speciation and Diversty. Academic Press, London.

Bartlett, J. K. and D. H. Vitt. 1986. A survey of species in the genus *Blindia* (Bryopsida, Seligeriaceae). New Zealand Journal of Botany 24: 203–246.

Bartram, E. B. 1931. A review of the American species of *Daltonia*. Bulletin of the Torrey Botanical Club 58: 31–48.

Bartram, E. B. 1934. Mosses of the River Napo, Ecuador. Revue Bryologique et Lichénologique 6: 9–18.

Bartram, E. B. 1949. Mosses of Guatemala. Fieldiana, Botany 25: 1–442.

Bartram, E. B. 1955. Mosses of the Ecuadorian Andes collected by P. R. Bell. Bulletin of the British Musum (Natural History), Botany 2: 51–64.

Bartram, E. B. 1964. Mosses of Cerro Antisana, Ecuadorian Andes. Revue Bryologique et Lichénologique 33: 1–14.

Bescherelle, E. 1894. Cryptogamae Centrali-Americanae in Guatemala, Costa-Rica, Columbia & Ecuador a cl. F. Lehmann lectae. Bulletin de l'Herbier Boissier 2: 389–400.

Bremer, B. 1980a. A taxonomic revision of *Schistidium* (Grimmiaceae, Bryophyta) 1. Lindbergia 6: 1–16.

Bremer, B. 1980b. A taxonomic revision of *Schistidium* (Grimmiaceae, Bryophyta) 2. Lindbergia 6: 89–117.

Bremer, B. 1981. A taxonomic revision of *Schistidium* (Grimmiaceae, Bryophyta) 3. Lindbergia 7: 73–90.

Brotherus, V. F. 1920 [1921]. Contributions à la flore bryologique de l'Ecuador. Revue Bryologique et Lichénologique 47: 1–16; 35–46.

Bruggeman-Nannenga, M.A. 1979. The section *Pachylomidium* (genus *Fissidens*) II. The species of Central America, temperate South America (including the High Andes), Australia, New Zealand and New Guinea. Proceeding of the Koninklijke Nederlandse Akademie van Wetenschappen, Series C, 82: 11–27.

Buck, W. R. 1979. A re-evaluation of the Bruchiaceae with the description of a new genus. Brittonia 31: 469–473.

Buck, W. R. 1980a. A generic revision of the Entodontaceae. Journal of the Hattori Botanical Laboratory 48: 71–159.

Buck, W. R. 1980b. Animadversions on *Pterigynandrum* with special commentary on *Forsstroemia* and *Leptopterigynandrum*. The Bryologist 83: 451–465.

Buck, W. R. 1983. A synopsis of the South American taxa of *Fabronia* (Fabroniaceae). Brittonia 35: 248–254.

Buck, W. R. 1984. *Bryosedgwickia*, a new synonym of *Platygyriella* (Hypnaceae). Brittonia 36: 86–88.

Buck, W. R. 1987a. Taxonomic and nomenclatural rearrangement in the Hookeriales with notes on West Indian taxa. Brittonia 39: 210–224.

Buck, W. R. 1987b. Notes on Asian Hypnaceae and associated taxa. Memoirs of the New York Botanical Garden 45: 519–527.

Buck, W. R. 1989. *Henicodium* replaces *Leucodontopsis* (Pterobryaceae). The Bryologist 92: 534.

Buck, W. R. 1991. Notes on neotropical Pterobryaceae. Brittonia 43: 96–101.

Buck, W. R. and H. Crum. 1978. A re-interpretation of the Fabroniaceae with notes on selected genera. Journal of the Hattori Botanical Laboratory 44: 347–369.

Buck, W. R. and H. Crum. 1990. An evaluation of familial limits among the genera traditionally aligned with the Thuidiaceae and Leskeaceae. Contributions from the University of Michigan Herbarium 17: 55–69.

Buck, W. R. and R. R. Ireland. 1985. A reclassification of the Plagiotheciaceae. Nova Hedwigia 41: 89–125.

Buck, W. R. and R. R. Ireland. 1989. Plagiotheciaceae. Flora Neotropica Monograph 50: 1–22.

Buck, W. R. and R. R. Ireland. 1992. *Symphyodon* (Symphyodontaceae) in the Americas. The Bryologist 95: 433–435.

Buck, W. R. and B. M. Thiers. 1989. Review of bryological studies in the tropics. Pp. 485–493. *In:* D. G. Campbell and H. D. Hammond (eds.), Floristic Inventory of Tropical Countries. New York Botanical Garden, Bronx, N.Y.

Burley, J. S. and N. M. Pritchard. 1990. Revision of the genus *Ceratodon* (Bryophyta). Harvard Papers in Botany 2: 17–76.

Bush, M. B., P. A. Colinvaux, M. C. Wiemann, D. R. Piperno, and K. Liu. 1990. Late Pleistocene temperature depression and vegetation change in Ecuadorian Amazonia. Quaternary Research 34: 330–345.

Churchill, S. P. 1981. A phylogenetic analysis, classification and synopsis of the genera of the Grimmiaceae (Musci). *In:* V. A. Funk and D. R. Brooks (eds.), Advances in Cladistics 1: 127–144.

Churchill, S. P. 1988a. Bryologia Novo Granatensis. Studies on the moss flora of Colombia I. New combinations, synonyms and comments. The Bryologist 91: 116–120.

Churchill, S. P. 1988b. A revision of the moss genus *Lepidopilum* (Callicostaceae). Ph. D. Dissertation, City University of New York.

Churchill, S. P. 1989. Bryologia Novo Granatensis. Estudios de los Musgos de Colombia IV. Catalogo nuevo de los Musgos de Colombia. Tropical Bryology 1: 95–132.

Churchill, S. P. 1990. *Pseudocrossidium steerei* (Pottiaceae), a new species from Ecuador. The Bryologist 93: 353–356.

Churchill, S. P. 1991a. Bryologia Novo Granatensis. V. Additional records for Colombia and Antioquia, with a review of the distribution of *Hydropogon fontinaloides* in South America. The Bryologist 94: 44–48.

Churchill, S. P. 1991b. The floristic compostion and elevational distribution of Colombian mosses. The Bryologist 94: 157–167.

Churchill, S. P. 1992. Clarification and review of *Lepidopilum affine* and *L. grevillianum* (Callicostaceae). Brittonia 44: 350–355.

Churchill, S. P. and W. R. Buck. 1982. A taxonomic investigation of *Leptotheca* (Rhizogoniaceae). Brittonia 34: 1–11.

Churchill, S. P. and E. Linares C. Prodromus Bryologiae Novo Granatensis. In preparation.

Churchill, S. P., I. Sastre-De Jesús and H. Balslev. 1992. The mosses of Añangu, Napo Province, Ecuador. Lindbergia 17: 50–54.

Cole, M. C. 1983. An illustrated guide to the genera of Costa Rican Hepaticae. I. Brenesia 21: 137–201.

Cole, M. C. 1984. Thallose liverworts and hornwarts of Costa Rica. Brenesia 22: 319–348.

Colinvaux, P. A. 1989. The past and future Amazon. Scientific American 260: 102–108.

Crosby, M. R. 1969. A revision of the tropical American moss genus *Pilotrichum*. The Bryologist 72: 275–343.

Crosby, M. R. 1974. Toward a revised classification of the Hookeriaceae (Musci). Journal of the Hattori Botanical Laboratory 38: 129–141.

Crosby, M. R. 1978. New combinations in *Callicosta* (Musci), the correct name for *Pilotrichum*. The Bryologist 81: 435–437.

Crosby, M. R., B. H. Allen and R. E. Magill. 1985. A review of the moss genus *Hypnella*. The Bryologist 88: 121–129.

Crum, H. A. 1957. A contribution to the moss flora of Ecuador. Svensk Botanisk Tidskrift 51: 197–206.

Crum, H. A. 1972. A taxonomic account of the Erpodiaceae. Nova Hedwigia 23: 201–224.

Crum, H. A. 1980. A guide to the identification of Mexican *Sphagna*. Contributions from the University of Michigan Herbarium 14: 25–52.

Crum, H. A. 1984a. Notes on tropical American mosses. The Bryologist 87: 203–216.

Crum, H. A. 1984b. Sphagnopsida. Sphagnaceae. North American Flora Series II, part 11: 1–180.

Crum, H. A. 1987a. New species of *Sphagnum* from South America. Journal of the Hattori Botanical Laboratory 63: 77–97.

Crum, H. A. 1987b. A new section and species of *Sphagnum* from Ecuador. Contributions from the University of Michigan Herbarium 16: 141–143.

Crum, H. A. 1989. Notes on South American species of *Sphagnum*. Journal of Bryology 15: 531–536.

Crum, H. A. 1990. A new look at *Sphagnum* sect. *Acutifolia* in South America. Contributions from the University of Michigan Herbarium 17: 83–91.

Crum, H. and D. Griffin, III. 1981. *Philonotis corticata*, new species from Mexico. The Bryologist 84: 399–401.

Deguchi, H. 1987. Studies on some Peruvian species of the Grimmiaceae (Musci, Bryophyta). Pp. 19–74. *In:* H. Inoue (ed.), Studies on Cryptogams in Southern Peru. Tokai University Press, Tokyo.

Delgadillo M., C. 1975. Taxonomic revision of *Aloina, Aloinella* and *Crossidium*. The Bryologist 78: 245–303.

Delgadillo M., C. 1982. Bryological exploration and moss research in the Neotropics. Pp. 507–512. *In:* P. Geissler and S. W. Greene (eds.), Bryophyte Taxonomy. J. Cramer.

De Luna, E. and W. R. Buck. 1991. An undescribed species of *Braunia* (Hedwigiaceae) from the Andean cloud forest. The Bryologist 94: 401–403.

De Notaris, J. 1859. Musci Napoani sive muscorum ad flumen Napo in Colombia a clar.mo Osculati lectorum. Memorie Della Reale Accademia Delle Scienze di Torino, Ser. II, 18: 437–455.

Eakin, D. A. 1976. A taxonomic revision of the moss genera *Rhegmatodon* and *Macrohymenium*. Ph. D. Dissertation. University of Florida.

Fife, A. J. 1987. Taxonomic and nomenclatural observations on the Funariaceae. 5. A revision of the Andean species of *Entosthodon*. Memoirs of the New York Botanical Garden 45: 301–325.

Florschütz, P. A. 1964. Musci, Part I, *In:* J. Lanjouw (ed.), Flora of Suriname. Leiden, Brill.

Florschütz-de Waard, J. 1986. Musci, Part II. *In:* A. L. Stoffers and J. C. Lindeman (eds.), Flora of Suriname. Leiden, Brill.

Florschütz-de Waard, J. 1990. A catalogue of the bryophytes of the Guianas. II. Musci. Tropical Bryology 3: 89–104.

Florschütz-de Waard, J. and J. M. Bekker. 1987. A comparative study of the bryophyte flora of different forest types in west Suriname. Cryptogamie, Bryologie et Lichénologie 8: 31–45.

Frahm, J.-P. 1978. Übersicht *Campylopus*-Arten der Anden. Journal of the Hattori Botanical Laboratory 44: 483–524.

Frahm, J.-P. 1981. Bestimmungsschlüssel und Illustrationen zu Gattung *Chorisodontium* Broth. Herzogia 5: 499–516.

Frahm, J.-P. 1983. A monograph of *Pilopogon* Brid. Lindbergia 9: 99–116.

Frahm, J.-P. 1984. A review of *Campylopodiella* Card. The Bryologist 87: 249–250.

Frahm, J.-P. 1987a. A revised list of the *Campylopus* species of the world. Bryologische Beitraege 7: 1–117.

Frahm, J.-P. 1987b. Which factors control the growth of epiphytic bryophytes in tropical rainforests? Symposia Biologica Hungarica 35: 639–648.

Frahm, J.-P. 1989. The genus *Chorisodontium* (Musci, Dicranceae) in the Neotropics. Tropical Bryology 1: 11–23.

Frahm, J.-P. 1991. Dicranaceae: Campylopodioideae, Paraleucobryoideae. Flora Neotropica Monograph 54: 1–238.

Fransén, S. 1988. On the status of *Bartramia campylopus* Schimp. in C. Müll. and *Gymnostomum setifolium* Hook. et Arnott. Lindbergia 14: 30–32.

Gentry, A. H. 1978. Floristic knowledge and needs in Pacific tropical America. Brittonia 30: 134–153.

Gentry, A. H. 1989. Speciation in tropical forests. Pp. 114–134, In: L. B. Holm-Nielsen, I. C. Nielsen and H. Balslev (eds.), Tropical forests: Botanical dynamics, speciation and diversty. Academic Press, London.

Gier, L. J. 1980. A preliminary study of the Thuidiaceae (Musci) of Latin America. Journal of Bryology 11: 253–309.

Giese, M. and J.-P. Frahm. 1985. A revision of *Microcampylopus* (C. Muell.) Fleisch. Lindbergia 11: 114–124.

Gradstein, S. R. and W. A. Weber. 1982. Bryogeography of the Galapagos Islands. Journal of the Hattori Botanical Laboratory 52: 127–152.

Griffin, D., III. 1970. Notes on the tropical genus *Prionodon*. Revue Bryologique et Lichénologique 37: 653–656.

Griffin, D., III. 1979. Guia preliminar para as briófitas freqüentes em Manaus e adjacências. Acta Amazonica 9 (Suplemento) : 1–57.

Griffin, D., III. 1981. El género *Sphagnum* en los Andes de Colombia y Venezuela. Cryptogamie, Bryologie et Lichénologie 2: 201–211.

Griffin, D., III. 1982. Los musgos del Estado Mérida: (clave para los géneros). Pittieria 11: 7–56.

Griffin, D., III. 1983. Keys to the genera of mosses from Costa Rica. Brenesia 21: 299–323.

Griffin, D., III. 1984a. A comparison of *Breutelia subarcuata* (C. Muell.) Schimp. in Besch. and *B. chrysea* (C. Muell.) Jaeg. in Latin America. The Bryologist 87: 233–237.

Griffin, D., III. 1984b. Studies on Colombian cryptogams XXII. The *Breutelia subarcuata* complex in Colombia and neighboring areas. Acta Botanica Neerlandica 3: 375–282.

Griffin, D., III. 1986a. *Oreoweisia* (Dicranaceae, Musci) in tropical America: An annotated key to species. Cryptogamie, Bryologie et Lichénologie 7: 433–438.

Griffin, D., III. 1986b. *Rhynchostegiella attenuata* Bartr. (Brachytheciaceae): a synonym of *Lindigia aciculata* (Tayl.) Hampe (Meteoriaceae). The Bryologist 89: 230–231.

Griffin, D., III. 1988. New World species of *Breutelia* with erect-appressed leaf bases. Beiheft zur Nova Hedwigia 90: 357–382.

Griffin, D., III. 1989a. Notes on morphological variation in *Breutelia inclinata* (Hamp. & Lor.) Jaeg. Journal of Bryology 15: 543–550.

Griffin, D., III. 1989b. *Oreoweisia erosa* (C. Muell.) Kindb., an african-neotropical disjunct. Cryptogamie, Bryologie et Lichénologie 10: 297–300.

Griffin, D., III, and W. R. Buck. 1989. Taxonomic and phylogenetic studies on the Bartramiaceae. The Bryologist 92: 368–380.

Griffin, D., III, and S. R. Gradstein. 1982. Bryological exploration of the tropical Andes: current status. Pp. 513–518. *In:* P. Geissler and S. W. Greene (eds.), Bryophyte Taxonomy. J. Cramer.

Grout, A. J. 1946. Bryales. Orthotrichaceae. North American Flora 15A, part 1: 1–62, Plates 1–5.

Grubb, P. J., J. R. Lloyd, T. D. Pennington and T. C. Whitmore. 1963. A comparison of montane and lowland rain forest in Ecuador I. The forest structure, physiognomy, and floristics. Journal of Ecology 51: 567–601.

Grubb, P. J. and T. C. Whitmore. 1966. A comparison of montane and lowland rain forest in Ecuador II. The climate and its effects on the distribution and physiognomy of the forests. Journal of Ecology 54: 303–333.

Grubb, P. J. and T. C. Whitmore. 1967. A comparison of montane and lowland rain forest in Ecuador III. The light reaching the ground vegetation. Journal of Ecology 55: 33–57.

Hampe, E. 1869. Musci frondosi a cl. H. Krause in Ecuador, Prov. Loja collecti. Botanische Zeitung 27: 433–437; 449–458.

Hegewald, E. 1978. Critical notes on *Holomitrium* (Dicranaceae) from the Antilles. The Bryologist 81: 524–531.

Horton, D. G. 1983. A revision of the Encalyptaceae (Musci), with particular reference to the North American taxa. Part II. Journal of the Hattori Botanical Laboratory 54: 353–532.

Hyvönen, J. 1989. A synopsis of genus *Pogonatum* (Polytrichaceae, Musci). Acta Botanica Fennica 138: 1–87.

Ireland, R. R. 1991. A preliminary study of the moss genus *Isopterygium* in Latin America. Caldasia 16: 265–276.

Koponen, A. 1977. *Tayloria* subgen. *Pseudotetraplodon*, subgen. nov., and new combinations in *Brachymitrion, Moseniella* and *Tayloria* (Splachnaceae, Musci). Annales Botanici Fennici 14: 193–196.

Koponen, A. 1981. Splachnobryaceae, a new moss family. Annales Botanici Fennici 18: 123–132.

Koponen, T. 1979. A synopsis of Mniaceae (Bryophyta, Musci) I. South and Central American taxa. Journal of the Hattori Botanical Laboratory 46: 155–161.

Lewinsky, J. 1976. On the systematic position of *Amphidium* Schimp. Lindbergia 3: 227–231.

Lewinsky, J. 1984. *Orthotrichum* Hedw. in South America 1. Introduction and taxonomic revision of taxa with immersed stomata. Lindbergia 10: 65–94.

Lewinsky, J. 1987. *Orthotrichum* (Orthotrichaceae) in South America 2. Taxonomic revision of taxa with superficial stomata. Memoirs of the New York Botanical Garden 45: 326–370.

Lewis, M. 1992. *Meteoridium* and *Zelometeorium* in Bolivia. Tropical Bryology 5: 35–53.

Løjtnant, B. and U. Molau. 1982. Analysis of a virgin páramo plant community on Volcán Sumaco, Ecuador. Nordic Journal of Botany 2: 567–574.

Long, D. G. 1979. A reassessment of the systematic position of *Tortula stanfordensis* Steere and *T. khartoumensis* Pettet. Journal of Bryology 10: 377–381.

Lin, S.-H. 1983. A taxonomic revision of Phyllogoniaceae (Bryopsida). Part I. Journal of the Taiwan Museum 36: 37–86.

Lin, S.-H. 1984. A taxonomic revision of Phyllogoniaceae (Bryopsida). Part II. Journal of the Taiwan Museum 37: 1–54.

Lorentz, P. G. 1868. Musci frondosi a clarissimo H. Krause in Ecuador, prov. Loja collecti. Botanische Zeitung 26: 793–800; 809–822.

Mägdefrau, K. 1982. Life-forms of bryophytes. Pp. 45–48. *In:* A. J. E. Smith (ed.), Bryophyte Ecology. London.

Malta, N. 1926. The genus *Zygodon*. Hook. et Tayl. Acta Horti Bot. Univ. Latvian 1: 1–184.

Manuel, M. 1977a. The genus *Meteoridium* (C. Müll.) Manuel, stat. nov. (Bryopsida: Meteoriaceae). Lindbergia 4: 45–55.

Manuel, M. 1977b. A monograph of the genus *Zelometeorium* Manuel gen. nov. Journal of the Hattori Botanical Laboratory 43: 107–126.

Manuel, M. 1977c. Studies in Cryphaeaceae IV. New combinations in *Schoenobryum*. The Bryologist 80: 522–524.

Manuel, M. 1981. Studies in Cryphaeaceae V. A revision of the family in Mexico, Central Amerca and the Caribbean. Journal of the Hattori Botanical Laboratory 49: 115–140.

Matteri, C. M. 1985. Current state of Latin American bryology. Journal of the Hattori Botanical Laboratory 59: 481–486.

McFarland, K. D. 1988. Revision of *Brachythecium* (Musci) for Mexico, Central America, South America, Antarctica, and Circum-subantarctic Islands. Ph. D. Dissertation, University of Tennessee, Knoxville.

Menzel, M. 1985[1986]. Die Gattung *Pogonatum* P. Beauv. (Polytrichales, Musci) in Lateinamerika 1. Taxonomie und Geographie von *Pogonatum campylocarpum* (C. Muell.) Mitt. und *P. subflexuosum* (Lor.) Broth. Lindbergia 11: 134–140.

Menzel, M. 1986. The genus *Pogonatum* P. Beauv. (Musci: Polytrichales) in Latin America. Lindbergia 12: 43–46.

Menzel, M. 1991. A taxonomic review of the genus *Lindigia* Hampe (Meteoriaceae, Leucodontales) and *Aerolindigia* gen. nov. (Brachytheciaceae, Hypnales), Bryopsida. Nova Hedwigia 52: 319–335.

Miller, H. A. 1971. An overview of the Hookeriales. Phytologia 21: 243–252.

Mitten, W. 1851. Catalogue of cryptogamic plants collected by Professor W. Jameson in the vicinity of Quito. Hooker's Journal of Botany 3: 49–57; 351–361.

Mitten, W. 1869. Musci Austro-Americani. Journal of the Linnean Society, Botany 12: 1–659.

Mohamed, M. R. 1979. A taxonomic study of *Bryum billardieri* Schwaegr. and related species. Journal of Bryology 10: 401–465.

Muñoz, L., H. Balslev and T. de Vries. 1985. Diversidad de la vegetación en cuatro cuadrantes en el páramo pajonal del Antisana, Ecuador. Publicaciones del Museo Ecuatoriano de Ciencias Naturales 6: 21–33.

Nishimura, N. 1985. A revision of the genus *Ctenidium* (Musci). Journal of the Hattori Botanical Laboratory 58: 1–82.

Nishimura, N. and H. Ando. 1986. A revision of some *Mittenothamnium* species described from Mexico. The Bryologist 89: 66–69.

Nishimura, N., M. Higuchi, T. Seki and H. Ando. 1984. Delimitation and subdivision of the moss family Hypnaceae. Journal of the Hattori Botanical Laboratory 55: 227–234.

Nyholm, E. 1971. Studies in the genus *Atrichum* P. Beauv. Lindbergia 1: 1–33.

Ochi, H. 1980. A revision of the neotropical Bryoideae, Musci (First part). Journal of the Faculty of Education, Tottori University, Natural Sciences 29: 49–154.

Ochi, H. 1981. A revision of the neotropical Bryoideae, Musci (Second part). Journal of the Faculty of Education, Tottori University, Natural Sciences 30: 21–55.

Ochyra, R. 1982. New names for genera of mosses. Journal of Bryology 12: 31–32.

Ochyra, R. 1991. *Hygrohypnum ellipticum* is *Bryum renauldii* (Musci). Fragmenta Floristica et Geobotanica 35: 71–75.

Ochyra, R., H. Bednarek-Ochyra, T. Pócs and M. R. Crosby. 1992. The moss *Adelothecium bogotense* in continental Africa, with a review of its world range. The Bryologist 95: 287–295.

Padberg, M. and J.-P. Frahm. 1985. Monographie der Gattung *Atractylocarpus* Mitt. (Dicranaceae). Cryptogamie, Bryologie et Lichénologie 6: 315–341.

Pócs, T. 1982. Tropical forest bryophytes. Pp. 59–104. *In:* A. J. E. Smith (ed.), Bryophyte Ecology. London.

Prance, G. T. 1984. Completing the inventory. Pp. 365–396. *In:* V. H. Heywood and D. M. Moore (eds.), Current Concepts in Plant Taxonomy. Academic Press, London.

Pursell, R. A. 1984. A preliminary study of the *Fissidens elegans* complex in the Neotropics. Journal of the Hattori Botanical Laboratory 55: 235–252.

Pursell, R. A. and D. M. Vital. 1986. Distributional adumbrations of *Fissidens* in the Neotropics. The Bryologist 89: 300–301.

Pursell, R. A., M. A. Bruggeman-Nannenga and B. H. Allen. 1988. A taxonomic revision of *Fissidens* subgenus *Sarawakia* (Bryopsida: Fissidentaceae). The Bryologist 91: 202–213.

Redfearn, P. L., Jr. 1991. *Gymnostomiella* (Musci: Pottiaceae) in the Neotropics and Eastern Asia. The Bryologist 94: 392–395.

Reese, W. D. 1961. The genus *Calymperes* in the Americas. The Bryologist 64: 89–140.

Reese, W. D. 1977. The genus *Syrrhopodon* in the Americas I. The elimbate species. The Bryologist 80: 1–31.

Reese, W. D. 1978. The genus *Syrrhopodon* in the Americas II. The limbate species. The Bryologist 81: 189–225.

Reese, W. D. 1979. New records of Calymperaceae in the Americas. Lindbergia 5: 96–98.

Reese, W. D. 1983. American *Calymperes* and *Syrrhopodon*: identification key and summary of recent nomenclatural changes. The Bryologist 86: 23–30.

Reese, W. D. 1987a. *Calymperes* (Musci: Calymperaceae): World ranges, implications for patterns of historical dispersion and speciation, and comments on phylogeny. Brittonia 39: 225–237.

Reese, W. D. 1987b. World ranges, implications for patterns of historical dispersal and speciation, and comments on phylogeny of *Syrrhopodon* (Calymperaceae). Memoirs of the New York Botanical Garden 45: 426–445.

Reese, W. D. 1993. Calymperaceae. Flora Neotropica Monograph, 58: 1–102.

Renner, S. S. 1993. A history of botanical exploration in Amazonia Ecuador, 1739–1988. Smithsonian Contributions to Botany, 82: 1–39.

Renner, S. S., H. Balslev and L. B. Holm-Nielsen. 1990. Flowering plants of Amazonian Ecuador - a checklist. AAU Reports 24: 1–241.

Richards, P. W. 1934. Musci collected by the Oxford expedition to British Guiana in 1929. Bulletin of Miscellaneous Information, Kew 1934: 317–337.

Richards, P. W. 1984. The ecology of tropical forest bryophytes. *In:* R. M. Schuster (ed.), New Manual of Bryology 2: 1233–1270. Nichinan.

Robinson, H. 1967. Preliminary studies on the bryophytes of Colombia. The Bryologist 70: 1–43.

Robinson, H. 1975. The mosses of Juan Fernandez Islands. Smithsonian Contributions to Botany 27: 1–88.

Robinson, H. 1987. Notes on generic concepts in the Brachytheciaceae and the new genus *Steerecleus*. Memoirs of the New York Botanical Garden 45: 678–681.

Robinson, H. and C. Delgadillo M. 1973. *Neosharpiella*, a new genus of Musci from high elevations in Mexico and South America. The Bryologist 76: 536–540.

Robinson, H., L. B. Holm-Nielsen and S. Jeppesen. 1971. Mosses of Ecuador. Lindbergia 1: 66–74.

Robinson, H., L. B. Holm-Nielsen and B. Løjtnant. 1977. Mosses of Ecuador II. Lindbergia 4: 105–116.

Salazar Allen, N. 1993. Leucophanaceae. Flora Neotropica Monograph 59: 1–11.

Salazar Allen, N. 1991. A preliminary treatment of the Central American species of *Octoblepharum* (Musci, Calymperaceae). Tropical Bryology 4: 85–97.

Salo, J. and M. Räsänen. 1989. Hierarchy of landscape patterns in western Amazon. Pp. 35–45. *In:* L. B. Holm-Nielsen, I. C. Nielsen and H. Balslev (eds.), Tropical Forests: Botanical Dynamics, Speciation and Diversty. Academic Press, London.

Sastre-De Jésus, I. 1987. A revision of the Neckeraceae Schimp. and Thamnobryaceae Marg. & Dur. in the Neotropics. Ph.D. Disseration, City University of New York.

Sastre-De Jésus, I. and S. P. Churchill. 1986. Manual de Briología Práctica: Asociación de Herbarios de Colombia. New York Botanical Garden. 24 pp.

Shaw, J. 1982. *Pohlia* Hedw. (Musci) in North and Central America and the West Indies. Contributions from the University of Michigan Herbarium 15: 219–295.

Shaw, J. 1984. Quantitative taxonomic study of morphology in *Epipterygium*. The Bryologist 87: 132–142.

Shaw, J. 1985. Nomenclatural changes in the Bryaceae, subfamily Mielichhoferioideae. The Bryologist 88: 28–30.

Shaw, J. 1987. Systematic studies on the Bryaceae. Memoirs of the New York Botanical Garden 45: 682–690.

Shaw, J. and H. Crum. 1982. Comments on the Mielichhoferioideae of Central America, with the description of a new species of *Synthetodontium*. Contributions from the University of Michigan Herbarium 15: 209–217.

Sharp, A. J., H. Crum and P. M. Eckel (eds.). In press. The Moss Flora of Mexico. Memoirs of the New York Botanical Garden.

Smith, G. L. 1971. A conspectus of the genera of Polytrichaceae. Memoirs of the New York Botanical Garden 21(3): 1–83.

Smith, G. L. 1975. Neotropical Polytrichaceae I, II. The Bryologist 78: 201–204.

Spruce, R. 1861. Mosses of the Amazon and Andes. Journal of the Linnean Society, Botany 5: 45–51.

Spruce, R. 1885. Hepaticae of the Amazon and the Andes of Peru and Ecuador. Transactions and Proceedings of the Botanical Scoiety. 15: i–xi, 1–588, tables I–XII. 1984 reprint with an introduction and index with updated nomenclature by B. M. Thiers in Contributions from the New York Botanical Garden, volume 15.

Spruce, R. 1908. Notes of a Botanist on the Amazon & Andes. Ed. A. R. Wallace. Volume II. MacMillan and Co., London.

Steere, W. C. 1948. Contributions to the bryogeography of Ecuador. I. A review of the species of Musci previously reported. The Bryologist 51: 65–167.

Steere, W. C. 1982. Four new species of Musci from the Andes of Ecuador and Colombia. Brittonia 34: 435–441.

Steere, W. C. 1986. *Trachyodontium*, a new genus of the Pottiaceae (Musci) from Ecuador. The Bryologist 89: 17–19.

Steere, W. C. 1988. Remarks on an ambiguous publication: Josepho De Notaris, Musci Naponani sive Muscorum ad Flumen Napo in Colombia a Clar: mo Osculati Lectorum. Beiheft zur Nova Hedwigia 90: 283–287.

Taylor, T. 1846. The distinctive characters of some new species of Musci, collected by Professor William Jameson, in the vicinity of Quito, and by Mr. James Drummond at Swan River. London Journal of Botany 5: 41–66.

Taylor, T. 1847. Descriptions of new Musci and Hepaticae, collected by Professor Williams Jameson on Pichincha, near Quito. London Journal of Botany 6: 328–342.

Taylor, T. 1848a. On some new Musci, collected by Professor W. Jameson on Pichincha. London Journal of Botany 7: 187–199.

Taylor, T. 1848b. On the specific characters of certain new cryptogamic plants, lately received from, and collected by, Professor William Jameson, on Pichincha, near Quito. London Journal of Botany 7: 278–285.

Thériot, I. 1936. Mousses de l'Equateur. Revue Bryologique et Lichénologique 9: 5–36.

Thiers, B. M. 1984. Introduction, pages ix–xi, *In:* R. Spruce, Hepaticae of the Amazon and the Andes of Peru and Ecuador. Contributions from the New York Botanical Garden. Volume 15.

Tixier, P. 1977. Clastobryoidées et taxa apparentés. Revue Bryologique et Lichénologique 43: 397–464.

Tixier, P. 1988. Le genre *Glossadelphus* Fleisch. (Sematophyllaceae, Musci) et sa valeur. Nova Hedwigia 46: 319–356.

Visnadi, S. R. and B. Allen. 1991. A revision of the genus *Lindigia* (Musci: Meteoriaceae). The Bryologist 94: 5–15.

Vitt, D. H. 1979. New taxa and new combinations in the Orthotrichaceae of Mexico. The Bryologist 82: 1–19.

Vitt, D. H. 1980. The genus *Macrocoma* I. Typification of names and taxonomy of the species. The Bryologist 83: 405–436.

Welch, W. H. 1943. The systematic position of the genera *Wardia, Hydropogon,* and *Hydropogonella*. The Bryologist 46: 25–46.

Welch, W. H. 1976. Hookeriaceae. North American Flora Series II, part 9: 1–133.

Whitmore, T. C. and G. T. Prance (eds.). 1987. Biogeography and Quarternary History in Tropical America. Clarendon Press, Oxford.

Wijk, R. van der, W. D. Margadant and P. A. Florschütz. 1959–1969. Index Muscorum I–V. Regnum Vegetabile 17, 26, 33, 48, 65.

Zander, R. H. 1972. Revision of the genus *Leptodontium* (Musci) in the New World. The Bryologist 75: 213–280.

Zander, R. H. 1977. The tribe Pleuroweisiae in Middle America. The Bryologist 80: 233–269.

Zander, R. H. 1978a. A synopsis of *Bryoerythrophyllum* and *Morinia* (Pottiaceae) in the New World. The Bryologist 81: 539–560.

Zander, R. H. 1978b. New combinations in *Didymodon* (Musci) and a key to the taxa in North America north of Mexico. Phytologia 41: 11–32.

Zander, R. H. 1979. Notes on *Barbula* and *Pseudocrossidium* (Bryopsida) in North America and an annotated key to the taxa. Phytologia 44: 177–214.

Zander, R. H. 1981a. Descriptions and illustrations of *Barbula, Pseudocrossidium* and *Bryoerythrophyllum* (p.p.) of Mexico. Cryptogamie, Bryologie et Lichénologie 2: 1–22.

Zander, R. H. 1981b. *Didymodon* (Pottiaceae) in Mexico and California: taxonomy and nomenclature of discontinuous and nondiscontinuous taxa. Cryptogamie, Bryologie et Lichénologie 2: 379–422.

Zander, R. H. 1983. Nomenclatural changes in *Hyophila, Leptodontium, Morinia* (Pottiaceae). The Bryologist 86: 156–157.

Zander, R. H. 1986. Notes on *Bryoerythrophyllum* (Musci). The Bryologist 89: 13–16.

Zander, R. H. 1989. Seven new genera in Pottiaceae (Musci) and a lectotype for *Syntrichia*. Phytologia 65: 424–436.

Zanten, B. O. van 1959. Trachypodiaceae. A critical revision. Blumea 9: 477–575.

Zomlefer, W. B. 1993. A revision of *Rigodium* (Musci: Rigodiaceae). The Bryologist 96: 1–72.

Zomlefer, W. B. and W. R. Buck. 1990. A reassessment of four *Rigodium* types. The Bryologist 93: 303–308.

REPORTS FROM THE BOTANICAL INSTITUTE,
UNIVERSITY OF AARHUS

1. **B. Riemann:** Studies on the Biomass of the Phytoplankton. 1976.
    Out of print.
2. **B. Løjtnant & E. Worsøe:** Foreløbig status over den danske flora. 1977.
    Out of print.
3. **A. Jensen & C. Helweg Ovesen (Eds.):** Drift og pleje af våde områder i
    de nordiske lande. 1977. 190 p. Out of print.
4. **B. Øllgaard & H. Balslev:** Report on the 3rd Danish Botanical
    Expedition to Ecuador. 1979. 141 p. Out of print.
5. **J. Brandbyge & E. Azanza:** Report on the 5th and 7th Danish-
    Ecuadorean Botanical Expeditions. 1982. 138 p.
6. **J. Jaramillo-A. & F. Coello-H.:** Reporte del Trabajo de Campo,
    Ecuador 1977—1981. 1982. 94 p.
7. **K. Andreasen, M. Søndergaard & H.-H. Schierup:** En karakteristik af
    forureningstilstanden i Søbygård Sø — samt en undersøgelse af
    forskellige restaureringsmetoders anvendelighed til en
    begrænsning af den interne belastning. 1984. 164 p.
8. **K. Henriksen (Ed.):** 12th Nordic Symposium on Sediments. 1984. 124
    p.
9. **L. B. Holm-Nielsen, B. Øllgaard & U. Molau (Eds.):** Scandinavian
    Botanical Research in Ecuador. 1984. 83 p.
10. **K. Larsen & P. J. Maudsley (Eds.):** Proceedings. First International
    Conference. European-Mediterranean Division of the
    international Association of Botanic Gardens. Nancy 1984. 1985. 90
    p.
11. **E. Bravo-Velasquez & H. Balslev:** Dinámica y adaptaciones de las
    plantas vasculares de dos ciénagas tropicales en Ecuador. 1985. 50
    p.
12. **P. Mena & H. Balslev:** Comparación entre la Vegetación de los
    Páramos y el Cinturón Afroalpino. 1986. 54 p.
13. **J. Brandbyge & L. B. Holm-Nielsen:** Reforestation of the High Andes
    with Local Species. 1986. 106 p.
14. **P. Frost-Olsen & L. B. Holm-Nielsen:** A Brief Introduction to the
    AAU - Flora of Ecuador Information System. 1986. 39 p.
15. **B. Øllgaard & U. Molau (Eds.):** Current Scandinavian Botanical
    Research in Ecuador. 1986. 86 p.
16. **J. E. Lawesson, H. Adsersen & P. Bentley:** An Updated and Annotated
    Check List of the Vascular Plants of the Galapagos Islands. 1987. 74
    p.
17. **K. Larsen:** Botany in Aarhus 1963 - 1988. 1988. 92 p.

AAU Reports:
18. Tropical Forests: Botanical Dynamics, Speciation, and Diversity.
    Abstracts of the AAU 25th Anniversary Symposium. Edited by **F.
    Skov & A. Barfod.** 1988. 46 pp.
19. Sahel Workshop 1989. University of Aarhus. Edited by **K. Tybirk, J. E.
    Lawesson & I. Nielsen.** 1989.

20. Sinopsis de las Palmeras de Bolivia. By **H. Balslev & M. Moraes.** 1989. 107 pp.

21. Nordiske Brombær (Rubus sect. Rubus, sect. Corylifolii og sect. sect. Caesii). By **A. Pedersen & J. C. Schou.** 1989. 216 pp.

22. Estudios Botánicos en la "Reserva ENDESA" Pichincha - Ecuador. Editado por **P. M. Jørgensen & C. Ulloa U.** 1989. 138 pp.

23. Ecuadorean Palms for Agroforestry. By **H. Borgtoft Pedersen & H. Balslev.** 1990. 120 pp

24. Flowering Plants of Amazonian Ecuador - a checklist. By **S. S. Renner, H. Balslev & L. B. Holm-Nielsen,** 1990. 220 pp.

25. Nordic Botanical Research in Andes and Western Amazonia. Edited by **S. Lægaard & F. Borchsenius,** 1990. 88 pp.

26. HyperTaxonomy - a computer tool for revisional work. By **F. Skov,** 1990. 75 pp.

27. Regeneration of Woody Legumes in Sahel. By **K. Tybirk,** 1991. 81 pp.

28. Régénération des Légumineuses ligneuses du Sahel. By **K. Tybirk,** 1991. 86 pp.

29. Sustainable Development in Sahel. Edited by **A. M. Lykke, K. Tybirk & A. Jørgensen,** 1992. 132 pp.

30. Arboles y Arbustos de los Andes del Ecuador. By **C. Ulloa Ulloa & P. M. Jørgensen,** 1992. 264 pp.

31. Neotropical Montane Forests. Biodiversity and Conservation. Abstracts from a Symposium held at The New York Botanical Garden, June 21–26, 1993. Edited by **Henrik Balslev,** 1993, 110 pp.

32. The Sahel. Population, Integrated Rural Development Projects, Research Components in Development Projects. Proceedings of the 5th Danish Sahel Workshop, 6-8 January 1994. Edited by **Anne'e Reenberg & Birgitte Markussen,** 1994. 171 pp.

33. The Vegetation of Delta de Saloum National Park, Senegal. By **Anne Mette Lykke,** 1994. 88 pp.

34. Seed Plants of the High Andes of Ecuador — a checklist. By **Peter M. Jørgensen & Carmen Ulloa Ulloa,** 1994. 443 pp.

35. The Mosses of Amazonian Ecuador. By **Steven P. Churchill,** 1994. 210 pp.

ORDER FORM

To Aarhus University Press
Aarhus University
DK-8000 Aarhus, DENMARK
Phone (+45) 8619 7033 Fax (+45) 8619 8433

I would like to order the following issues of *AAU Reports:*

_____

Do not send payment with your order. We will bill you later

Please send the books to:

Name:
Street:
Town:
Country: